Raising Beef Cattle FOR DUMMIES®

by Nikki Royer and Scott Royer

John Wiley & Sons, Inc.

Raising Beef Cattle For Dummies®

Published by
John Wiley & Sons, Inc.
111 River St.
Hoboken, NJ 07030-5774
www.wiley.com

Copyright © 2012 by John Wiley & Sons, Inc., Hoboken, New Jersey

Published simultaneously in Canada

No part of this publication may be reproduced, stored in a retrieval system or transmitted in any form or by any means, electronic, mechanical, photocopying, recording, scanning or otherwise, except as permitted under Sections 107 or 108 of the 1976 United States Copyright Act, without the prior written permission of the Publisher. Requests to the Publisher for permission should be addressed to the Permissions Department, John Wiley & Sons, Inc., 111 River Street, Hoboken, NJ 07030, (201) 748-6011, fax (201) 748-6008, or online at http://www.wiley.com/go/permissions.

Trademarks: Wiley, the Wiley logo, For Dummies, the Dummies Man logo, A Reference for the Rest of Us!, The Dummies Way, Dummies Daily, The Fun and Easy Way, Dummies.com, Making Everything Easier, and related trade dress are trademarks or registered trademarks of John Wiley & Sons, Inc., and/or its affiliates in the United States and other countries, and may not be used without written permission. All other trademarks are the property of their respective owners. John Wiley & Sons, Inc., is not associated with any product or vendor mentioned in this book.

LIMIT OF LIABILITY/DISCLAIMER OF WARRANTY: THE PUBLISHER AND THE AUTHOR MAKE NO REPRESENTATIONS OR WARRANTIES WITH RESPECT TO THE ACCURACY OR COMPLETENESS OF THE CONTENTS OF THIS WORK AND SPECIFICALLY DISCLAIM ALL WARRANTIES, INCLUDING WITHOUT LIMITATION WARRANTIES OF FITNESS FOR A PARTICULAR PURPOSE. NO WARRANTY MAY BE CREATED OR EXTENDED BY SALES OR PROMOTIONAL MATERIALS. THE ADVICE AND STRATEGIES CONTAINED HEREIN MAY NOT BE SUITABLE FOR EVERY SITUATION. THIS WORK IS SOLD WITH THE UNDERSTANDING THAT THE PUBLISHER IS NOT ENGAGED IN RENDERING LEGAL, ACCOUNTING, OR OTHER PROFESSIONAL SERVICES. IF PROFESSIONAL ASSISTANCE IS REQUIRED, THE SERVICES OF A COMPETENT PROFESSIONAL PERSON SHOULD BE SOUGHT. NEITHER THE PUBLISHER NOR THE AUTHOR SHALL BE LIABLE FOR DAMAGES ARISING HEREFROM. THE FACT THAT AN ORGANIZATION OR WEBSITE IS REFERRED TO IN THIS WORK AS A CITATION AND/OR A POTENTIAL SOURCE OF FURTHER INFORMATION DOES NOT MEAN THAT THE AUTHOR OR THE PUBLISHER ENDORSES THE INFORMATION THE ORGANIZATION OR WEBSITE MAY PROVIDE OR RECOMMENDATIONS IT MAY MAKE. FURTHER, READERS SHOULD BE AWARE THAT INTERNET WEBSITES LISTED IN THIS WORK MAY HAVE CHANGED OR DISAPPEARED BETWEEN WHEN THIS WORK WAS WRITTEN AND WHEN IT IS READ.

For general information on our other products and services, please contact our Customer Care Department within the U.S. at 877-762-2974, outside the U.S. at 317-572-3993, or fax 317-572-4002.

For technical support, please visit www.wiley.com/techsupport.

Wiley publishes in a variety of print and electronic formats and by print-on-demand. Some material included with standard print versions of this book may not be included in e-books or in print-on-demand. If this book refers to media such as a CD or DVD that is not included in the version you purchased, you may download this material at http://booksupport.wiley.com. For more information about Wiley products, visit www.wiley.com.

Library of Congress Control Number is available from the Publisher upon request.

ISBN 978-0-470-93061-8 (pbk); ISBN 978-1-118-08911-8 (ebk); ISBN 978-1-118-08912-5 (ebk); ISBN 978-1-118-08913-2 (ebk)

Manufactured in the United States of America

10 9 8 7 6 5 4 3 2 1

About the Authors

Scott and Nikki Royer raise cattle, sheep, pigs, and chickens on their diversified family farm in Indiana. In addition to raising the calves from their own cow herd, the Royers also purchase and finish feeder cattle from other producers. Nikki has been involved with raising cattle her entire life, showing her first calf at the age of six. For many years she and her family exhibited champion cattle across the country. Both Scott and Nikki are graduates of Purdue University School of Agriculture. Scott worked in the animal feed business and as an animal health research scientist before becoming a full-time livestock producer in 2001. After receiving a master's degree in Meat Science, Nikki was a sales representative until joining Scott to raise cattle and kids on the farm that her family started in 1874.

Scott and Nikki's mission at their farm is to provide delicious, nutritious food that's raised sustainably in a low-stress pasture environment. In 2010, they direct marketed over 45,000 pounds of meat through farmer's markets, on-farm sales, and the Internet.

Dedication

Nikki Royer: I dedicate this book to my dad, Knic Overpeck, who was an amazing cattleman.

Author's Acknowledgments

Nikki Royer: Thanks to my husband Scott for being the best spouse and partner I could ever hope to have. To the rest of my family, thank you for your patience during this project. I would like to acknowledge Stephanie for the recommendation and Jennifer Moore, Jessica Smith, David Lutton, and Erin Calligan Mooney for guiding the creation of this book.

Publisher's Acknowledgments

We're proud of this book; please send us your comments at http://dummies.custhelp.com. For other comments, please contact our Customer Care Department within the U.S. at 877-762-2974, outside the U.S. at 317-572-3993, or fax 317-572-4002.

Some of the people who helped bring this book to market include the following:

Acquisitions, Editorial, and Vertical Websites

Project Editor: Jennifer Moore

Acquisitions Editors: Erin Calligan Mooney

Copy Editor: Jessica Smith

Assistant Editor: David Lutton

Editorial Program Coordinator: Joe Niesen

Technical Editor: Mark Castor

Editorial Manager: Carmen Krikorian

Senior Editorial Manager: Jennifer Ehrlich

Editorial Assistants: Alexa Koschier, Rachelle Amick

Art Coordinator: Alicia B. South

Cover Photos: © iStockphoto.com/ Michael Gray

Cartoons: Rich Tennant (www.the5thwave.com)

Composition Services

Project Coordinator: Patrick Redmond

Layout and Graphics: Jennifer Creasey, Melanee Habig, Joyce Haughey

Proofreaders: Rebecca Denoncour, Melanie Hoffman, Jessica Kramer, Lauren Mandelbaum, Dwight Ramsey

Indexer: Christine Karpeles

Illustrator: Barbara Frake

Special Help
Christine Pingleton

Publishing and Editorial for Consumer Dummies

Kathleen Nebenhaus, Vice President and Executive Publisher

Kristin Ferguson-Wagstaffe, Product Development Director

Ensley Eikenburg, Associate Publisher, Travel

Kelly Regan, Editorial Director, Travel

Publishing for Technology Dummies

Andy Cummings, Vice President and Publisher

Composition Services

Debbie Stailey, Director of Composition Services

Contents at a Glance

Introduction 1

Part I: Getting to Know Beef Cattle 7

Chapter 1: The Rewards and Responsibilities of Raising Beef Cattle 9

Chapter 2: Bovine Basics: Understanding Cattle Terminology and Anatomy 23

Chapter 3: From Angus to Zebu: Looking at Beef Cattle Breeds and Traits 35

Part II: Preparing to Bring Home the Beef 53

Chapter 4: Home on the Range: Preparing Your Facilities 55

Chapter 5: Chow Time: Planning Your Feeding Program 79

Chapter 6: Exploring Pasture, Manure, and Water Management 107

Chapter 7: Choosing and Buying Cattle 127

Part III: Cattle Handling, Health, and Breeding 145

Chapter 8: Understanding and Properly Handling Cattle 147

Chapter 9: Keeping Your Cattle Healthy 165

Chapter 10: Addressing Common Cattle Ailments 191

Chapter 11: Breeding Cows and Caring for Pregnant Females 213

Chapter 12: Looking After Calves Young and Old 241

Part IV: Realizing Your Cattle Business Potential 261

Chapter 13: Turning Your Extra Pastures into a Money-Making Business 263

Chapter 14: Showing and Selling Cattle 275

Chapter 15: Managing Your Beef Business 293

Part V: The Part of Tens 311

Chapter 16: Ten Common Mistakes to Avoid When Raising Cattle 313

Chapter 17: Ten (or so) Bizarre Bovine Behaviors . . . and What They Mean 321

Index 327

Table of Contents

Introduction 1

About This Book 1
Conventions Used in This Book 2
What You're Not to Read 2
Foolish Assumptions 3
How This Book Is Organized 3
Part I: Getting to Know Beef Cattle 3
Part II: Preparing to Bring Home the Beef 4
Part III: Cattle Handling, Health, and Breeding 4
Part IV: Realizing Your Cattle Business Potential 4
Part V: The Part of Tens 4
Icons Used in This Book 5
Where to Go from Here 5

Part I: Getting to Know Beef Cattle 7

Chapter 1: The Rewards and Responsibilities of Raising Beef Cattle 9

Burgers and Beyond: Examining the Benefits of Owning Beef Cattle 9
Becoming more self-sufficient 10
Producing extra income 10
Improving the land naturally 11
Providing a fun and educational family project 11
Providing the Basics for Your Cattle 12
Wholesome feed and water 12
Clean, spacious living and eating areas 12
A low-stress environment 13
Planning for the Labor and Financial Commitments 14
Reviewing daily chore commitments 14
Planning for big projects and emergencies 14
Looking at purchase prices 15
Calculating feed expenses 15
Figuring the cost of healthcare 16
Budgeting for facilities and land 17
Developing a Herd Mentality: Assembling Your Team of Experts 17
Making friends with a mentor 17
Vetting your veterinarian 18
Relying on local extension agents 18
Asking for advice from your feed dealer 19

Holy Cow! Important Safety and Legal Considerations 19
Preventing physical injuries 19
Watching out for zoonotic diseases 20
Considering your insurance needs 21
Being a good neighbor 21
Complying with zoning regulations 22

Chapter 2: Bovine Basics: Understanding Cattle Terminology and Anatomy 23

Livestock Lingo: Describing Cattle by Age, Sex, and Life Stage 23
Grasping the Wonders of the Bovine Body 25
"Nice pin bones": Getting acquainted with general cattle anatomy 25
Ruminating on cattle stomachs 27
Checking out male reproductive anatomy 29
Making sense of female reproductive anatomy 30
Meat or Milk? Investigating Different Types of Cattle 31
Looking at the udderly fascinating world of dairy cattle 32
The big beef: Eyeing the traits of beef cattle 33

Chapter 3: From Angus to Zebu: Looking at Beef Cattle Breeds and Traits 35

Identifying the Traits That Are Important to You 35
Keeping up with the Jones's: Follow your neighbors' lead when choosing breeds 36
Purple cows need not apply: Picking your cattle's hair color 37
Horning in: Choosing animals with or without horns 38
Matching breed characteristics to your feed resources and customer needs 39
Perusing breed purity: Purebred or crossbred 40
Beefing Up Your Knowledge of Common Beef Breeds 42
Introducing the British breeds 43
A well-cultured cow: Looking at the Continental breeds 46
The American Brahman: The best known North American Bos indicus 50
Maintaining vigor with the composite breeds 50
Heritage breeds: Examining the old, pure bloodlines 51

Part II: Preparing to Bring Home the Beef 53

Chapter 4: Home on the Range: Preparing Your Facilities 55

Making (and Keeping) Good Neighbors by Building Strong Fences 55
Addressing permanent fence needs 56
Preparing internal fencing 60
Planning for gates 60
Keeping cattle in line with electric fencing 62

Planning Your Handling Facilities 65
Selecting a site for your handling facilities 65
Designing your handling facilities 66
Selecting a head gate and holding chute 69
Sheltering Your Cattle 70
Going with the tried and true: Pole barns 71
Checking out the latest in barnyard architecture: Hoop barns 72
Making do with what you have: Repurposing existing structures 72
Providing shade sources and wind breaks 73
Choosing bedding for your cattle shelter 75
Preventing Mud and Muck with a HUAP Zone 76

Chapter 5: Chow Time: Planning Your Feeding Program 79

Understanding the Nutrients Cattle Need 80
Starting with protein 80
Harnessing the nutritional power of energy: Carbs, fats, and proteins 83
Mighty minerals: Finding the right balance 84
Cooking up an alphabet soup of vital vitamins 89
Water: Providing the ever-important liquid 91
Comparing Feeds and Choosing the Best for Your Cattle 92
Feeding forages 93
Adding to the cattle diet with concentrates 94
Considering nontraditional feedstuffs 95
Doling Out the Proper Feed for All Types of Cattle 95
Nourishing the newborn 96
Providing for the growing calf 97
Balancing the diet for a finishing animal 98
Maintaining the dry cow 98
Keeping your pregnant or lactating cows fed 99
Meeting the dietary needs of your bull 99
Deciding on a Feeding Schedule 100
Opting for limit feeding 100
Allowing cattle to free-choice feed 101
Introducing new feeds to the diet 101
Purchasing the Basic Supplies: Fine China Not Required 102
Getting excited about grain feeders 102
Relying on hay rings 104
Ensuring proper water sources 104

Chapter 6: Exploring Pasture, Manure, and Water Management . . .107

Pasture Basics: The Forage Elements that Make for Good Grazing 108
Characterizing quality forages 108
Looking at the types of forages that are available for your pastures 109
Seeking Greener Pastures: Improving or Establishing Your Grazing Land 111

Evaluating and perfecting an existing pasture 112
Preparing a new pasture 113
Improving your pasture by fertilizing and feeding the soil 114
Making manure a nutrient for your pasture, not a nuisance 115
Optimizing Pasture Productivity 118
Matching the number of cattle to the amount of available pasture 119
Giving your pastures a rest with rotational grazing 121
Creating a yearlong forage chain 122
Protecting Your Water Resources 124
Building a stream crossing and drinking area 124
Managing your ponds and cattle together 125
Protecting riparian areas with flash grazing 126

Chapter 7: Choosing and Buying Cattle 127

Selecting the Right Cattle for You 127
Early start: Considering the fun and commitment of a bottle calf 128
Ready to grow: Exploring the benefits of a weaned calf 130
Blue ribbons: Enjoying the rewards of a show calf 130
A bun in the oven: Pondering the promise of a bred cow 131
Deciding Where to Buy Your Cattle: A Bovine Shopper's Paradise 132
Going to the farm 133
Checking out consignment sales 134
Taking advantage of test stations 135
Visiting the local sale barn 135
Surfing the Internet 136
Making a Wise Purchase: What to Look For Before You Buy 136
Noticing signs of a healthy versus unhealthy animal 137
Avoiding genetic and bloodborne diseases 138
Assessing animal disposition 138
Judging conformation 139
Understanding performance measurements and pedigrees 140
Evaluating customer service 141
Arriving Home and Settling In 142
Transporting your cattle 142
Adjusting cattle to their surroundings and feedstuffs 143

Part III: Cattle Handling, Health, and Breeding 145

Chapter 8: Understanding and Properly Handling Cattle 147

Getting Acquainted with Bovine Behavior 148
Recognizing cattle preferences 148
Expecting cattle fighting 149
Dealing with riding or mounting cattle 150

Determining Whether Your Cattle Are Content 150
Deciphering body language: The head, the tail, and the stuff in between 151
Translating the moo 153
Taking Advantage of Your Cattle's Natural Behaviors 154
Managing the predator-prey relationship 154
Using the flight zone to move cattle 155
Working with cattle in a handling facility 158
Training your cattle 160
Safely Handling Breeding Stock 160
Selecting the right equipment 161
Evaluating a bull's attitude and handling him 161
Working with mothers and baby calves 163

Chapter 9: Keeping Your Cattle Healthy........165

Observing Your Cattle to Recognize Illness Early 165
Keeping an eye on appetite and behavior 166
Checking urine and feces 167
Get His Vitals, Stat: Measuring Physical Vital Signs 168
Checking body temperature 168
Taking the pulse 170
Observing respiration 171
Deciding what to do for a sick animal 171
Vaccinating Your Animals 174
Creating an Annual Healthcare Calendar 175
Starting in pre-calving and calving season 175
Moving on to breeding season 176
Entering into the preweaning and weaning season 177
Preventing Seasonal Diseases and Issues 178
Being a bad host to internal and external parasites 178
Controlling pinkeye 181
Beefing up seasonal nutritional needs 181
Keeping cattle comfortable during extreme weather 182
Mastering Basic Animal Husbandry Skills 183
First things first: Stocking your first-aid kit 183
Making injections a breeze 185
Hi! My name is Daisy: Attaching identifying ear tags 187
Treating your cattle's wounds 189

Chapter 10: Addressing Common Cattle Ailments........191

Understanding Respiratory Diseases 191
Infectious bovine rhinotracheitis (IBR), or red nose 194
Bovine virus diarrhea (BVD) 194
Bovine respiratory syncytial virus (BRSV) 195
Parainfluenza-3 (PI-3) 195
Pasteurella 196
Haemophilus somnus (H. somnus) 196

Closing In on Clostridial Diseases 196
Dealing with Reproductive Diseases 199
Brucellosis 199
Leptospirosis 200
Vibriosis 200
Trichomoniasis 201
Contending with Other Infectious Diseases 201
Pinkeye 201
Johne's disease 202
Tuberculosis 203
Scours 203
Looking Out for External Parasites 204
Flies 204
Grubs 206
Lice 207
Ringworm 207
Mites 208
Controlling Internal Parasites 209
Preventing Feed-Related Problems 210
Moldy feed 210
Poisonous plants 210
Founder 211
Bloat 211
Sweet clover poisoning 212

Chapter 11: Breeding Cows and Caring for Pregnant Females 213

Preparing Your Cow for a Successful Pregnancy 213
Assessing body condition scores 214
Providing extra preparations for cows at either end of the age spectrum 217
Administering reproductive vaccines 219
Determining a due date after conception 220
Tending to the Male Side of the Reproductive Equation 221
Selecting the right bull for your cows 221
Understanding the breeding soundness exam 223
Caring for your bull 224
Getting Cows Pregnant without Owning a Bull 225
Opting for artificial insemination 225
Leasing a bull 229
Planning for the Big Delivery 230
Setting up the birthing spaces 230
Gathering your calving supplies 231
Monitoring the Stages of Labor 233
Gearing up for pre-delivery 233
Progressing into active labor 233
Watching for postpartum actions 237
Managing Potential Issues during Pregnancy and after Delivery 237
Coping with abortions 237
Being aware of milk fever 239

Dealing with temporary paralysis........239
Preventing udder issues........240

Chapter 12: Looking After Calves Young and Old........241

Welcome, Baby: Caring for a New Bovine Arrival........242
Helping the calf with breathing difficulties........242
Treating the navel........243
Where's my name tag? Identifying the calves........243
Ensuring the calf gets enough colostrum........246
Dodging Calf Problems in the First Few Days........249
Brrrr: Dealing with hypothermia........250
Diagnosing navel or joint ill........251
Looking out for scours........252
Growing Up: Tending to the Older Calf........254
Expanding the calf cuisine: Providing creep feed........254
Developing a vaccination schedule........255
Handling horns........256
Castrating young bulls........257
Easing the stress of weaning........258

Part IV: Realizing Your Cattle Business Potential....... 261

Chapter 13: Turning Your Extra Pastures into a Money-Making Business........263

Growing Stocker Calves........263
Deciding on a season........264
Identifying a source........265
Selecting good stocker calves........265
Planning for a buyer........267
Preparing your fields, facilities, and cattle........267
Tending Cows for Absentee Owners........268
Matching the number of cows on your wish list to your resources........269
Finding cattle (and their owners)........270
Creating a contract for services........270
Raising 100 Percent Grass-Fed Beef........271
Locating customers........272
Supplying proper pasture feedstuffs........273
Determining what type of cattle to raise on an all-grass diet........273

Chapter 14: Showing and Selling Cattle........275

Preparing Yourself and Elsie for an Exhibition........275
Setting your goals........276
Finding the right show for you........277
Training your animal to be well-mannered in the show ring........280
Packing your feeding and grooming supplies for the show........283

Making the Most of the Show 284
Putting your best foot (or hoof) forward 285
Shave and a haircut, two bits: Grooming your animal 285
Following show ring protocol 287
Exploring Sales Opportunities for Your Cattle 289
Opting for private treaty deals 289
Participating in consignment sales 290
Getting a hoof in the door by enrolling in a performance test 290
Selling your cattle online 291

Chapter 15: Managing Your Beef Business 293

Covering the Legalities of Selling Beef 293
Understanding governmental rules regarding processing facilities 294
Obtaining permits and licenses 295
Protecting yourself with proper insurance coverage 296
Exploring and Pricing the Meat a Market Animal Produces 296
Selling the whole animal on the hoof 297
Providing freezer beef 297
Marketing your meat by the individual cut 298
Setting your prices 300
Finding and Working with Your Butcher 302
Checking out the facilities and personnel 303
Considering your animals' well-being 304
Deciding on packages and labels 304
Maintaining a good relationship with your chosen meat cutter 305
Exploring Your Different Selling Options 306
Selling direct off the farm 306
Participating in farmers' markets 306
Harnessing the power of the Internet 307
Working with restaurants 307
Promoting and Marketing Your Beef 308
Beefing up on cattle-selling terminology 308
Spreading the word about your products 309

Part V: The Part of Tens 311

Chapter 16: Ten Common Mistakes to Avoid When Raising Cattle 313

Tolerating Weak Perimeter Fences 313
Running Out of Grass 314
Forgoing a Calving Ease Bull 315
Neglecting to Keep an Adequate Backup Water Supply 315

Lacking a Marketing Plan....316
Buying Bargain Cattle....316
Purchasing Cattle for Looks and Not Function....317
Skimping on Training before the Cattle Show....318
Failing to Set Proper Customer Expectations....318
Making Raising Cattle Complicated....319

Chapter 17: Ten (or so) Bizarre Bovine Behaviors . . . and What They Mean....321

Eating Old, Brown Grass When Fresh Forage Is Available....321
Squirting Projectile Manure....322
Not Drinking from the Water Trough....322
Galloping around the Pasture....323
Pacing along the Fence....324
Balking at Shadows or Holes....324
Bawling Frantically....325
Rub-a-Dub-Dubbing on Anything and Everything....325
Curling Their Lips....326

Index....327

Introduction

We've had the honor and pleasure of caring for and working with cattle for more than three decades. In the past few years, we've been delighted to see more folks joining the cattle raising tradition as new herds pop up across the country. We've seen it all. Some folks want to raise an animal or two to keep their own freezers stocked with beef. Others are looking to make a little extra money by growing cattle on extra grass instead of having to mow it. And still others are butchering their animals and selling the meat directly to customers. Regardless of your reason for raising cattle, taking care of these amazing creatures is a great educational experience that the whole family can be involved in.

In this book, we share some of our hard-earned knowledge and ideas so you can better care for your cattle right now instead of relying solely on the tough lessons taught at the "school of experience." So whether you're looking to raise livestock for the first time, thinking about adding cattle to your current farm-animal mix, or seeking answers to some questions about the cattle you already own, this book is for you.

About This Book

So many bits and pieces of information regarding raising cattle float around, particularly online, so it can be hard to figure out what is essential and what is mere window dressing. In this book, we bring all the essential aspects of cattle production together.

We cover the gamut of the wonders and responsibilities that await you in the bovine world. We give you basic information on how to find the best cattle, including descriptions of breeds, types of cattle, and places to buy them. And we provide you with all you need to know to get started housing and feeding your herd.

Of course, you also want to keep your animals healthy and content, so we provide you with a preventive health plan and clue you in to the possible illnesses your cattle may encounter and how to treat them. We also delve into many other aspects of cattle production: taking care of pregnant cows and baby calves, showing your animals, marketing your beef, and many more.

We provide modern, research-based information that applies to and works in the real world. Many of the tips and techniques we talk about come from our own personal experience.

The best part is that you don't have to read this book from cover to cover to get started on your cattle journey. Pick a page, any page, and then start reading. This book is set up in a modular fashion so you can jump around to suit your particular needs. Dabble in whatever chapters you need right now. However, if you want to get an all-encompassing overview of the aspects of raising cattle, you're more than welcome to read each and every word we write in order from front cover to back.

Conventions Used in This Book

We've adopted the following conventions throughout this book to make the material easy to access and understand:

- New or unfamiliar terms appear in *italics* and are accompanied by a concise definition.
- **Bold** indicates the action to take in numbered steps and highlights the key words in bulleted lists.
- Web addresses are set in `monofont` so you can easily spot them.

When this book was printed, some web addresses may have needed to break across two lines of text. If that happened, rest assured that we haven't put in any extra characters (such as hyphens) to indicate the break. So when using one of these web addresses, just type in exactly what you see in this book, pretending as though the line break doesn't exist.

What You're Not to Read

All the pages in this book are chock-full of good info, but you don't have to read every line to understand the topic at hand (unless you just want to). For instance, text preceded by the Technical Stuff icon gives you in-depth background information that isn't necessary but that is nonetheless interesting. You won't be arrested for cattle neglect if you skip this material. The same holds true for the gray-shaded sidebars, which contain fun and fascinating (at least we hope!) but nonessential information.

Foolish Assumptions

Because a book that contains everything there is to know about cattle would probably weigh as much as a mature cow (around 1,100 pounds, in case you're wondering), we made some assumptions about you, the reader, and what you would want to know. Here are the assumptions we made about you:

- You have some general experience taking care of animals, but you aren't an expert on cattle and need some basic information.
- You want to take proper care of your cattle so they stay healthy and content.
- You realize raising cattle isn't an inexpensive endeavor, and you don't mind spending a little (or a lot) of money so you can do it right the first time.
- You want to raise cattle for their meat rather than for their milk (if you need help understanding the difference between beef and dairy cattle, flip to Chapter 2).

How This Book Is Organized

To keep this book easy to use, we've organized it in parts by general topics. The chapters help narrow the focus to a particular area of raising cattle, such as caring for a pregnant cow or tending to a calf. We think you'll use this book as a go-to reference all the time. The following sections provide you with a preview of what you can expect from the parts in this book.

Part I: Getting to Know Beef Cattle

This part looks at the benefits and potential concerns associated with raising cattle. It also introduces the vocabulary that goes along with having cattle. We round out the part by covering some basic cattle welfare information and exploring the broad array of cattle breeds and their various characteristics.

Part II: Preparing to Bring Home the Beef

Taking the time to prepare your facilities and plan your feeding program can help make raising cattle more enjoyable and hopefully more profitable. The chapters in this part help you with the steps needed to prepare your pastures, fences, and corrals. We also take a look at what your cattle like (and need) to eat and drink. After you've prepared for bringing home some cattle, you're ready to go find and buy the animals. We show you how to match your prospective cattle to your specific needs.

Part III: Cattle Handling, Health, and Breeding

In this part, you can find an overview of working with and handling cattle and tending to their preventive healthcare. We also provide information on common cattle diseases and injuries and their treatments. The part wraps up with time-tested recommendations for caring for cattle at different stages of the life cycle.

Part IV: Realizing Your Cattle Business Potential

Cattle are not only fun to raise, but they're also often profitable as well. This part explores three options for your cattle business: using pastures and grazing, showing and selling cattle, and selling beef direct to customers.

Part V: The Part of Tens

This final part is a *For Dummies* hallmark. The Part of Tens gives you quick lists of tips. For our Part of Tens, we introduce ten cattle-raising mistakes to avoid and ten bovine behaviors to understand and translate. After all, your cattle will try to communicate with you!

Icons Used in This Book

Throughout the book, you'll come across little pictures in the left margins. These pictures, called icons, can help you classify the information found beside them. Here are the icons we use in this book:

These bits of text include ideas or pointers that help you save time, money, or frustration (and maybe all three!).

The information beside this icon is important. You want to read this stuff closely so you can keep your cattle and yourself safe, content, and profitable.

When you see this icon, pay attention. This red flag highlights information that can help protect you and your cattle from dangerous situations.

If you like to know all the details, this icon is for you. But, if you just want the basics, feel free to skip these paragraphs.

Where to Go from Here

If you're ready to get started down the path to bovine bliss, just determine what you need to know and head to the chapter that discusses that topic. You can flip from chapter to chapter without missing a beat. Or, if you prefer, start from the beginning to get an overview. The choice is yours.

Here are some suggestions of some specific bovine educational needs you may have:

- Check out Chapter 4 for ideas on how to safely contain your cattle.
- If you're overwhelmed when it comes to planning the feed ration for your cattle, check out Chapter 5.
- If your cattle seem ill, go to Chapter 10 for a review of common ailments and effective treatments.
- Start with Chapter 14 if you're preparing to exhibit cattle for the first time.

And don't forget that the in-depth table of contents and index can help you find the information you need, and clear, eye-catching headings direct you right where you want to go. Best of luck with your bovine journey; we hope you enjoy working with cattle as much as we do!

Part I
Getting to Know Beef Cattle

"So, you think that's the cow that ate the steroid-laced feed?"

In this part . . .

The chapters in this part help you make well-informed decisions about raising beef cattle. Chapter 1 takes a look at what's involved in the big job of raising cattle. Chapter 2 gets you up to speed on the latest livestock language and terminology, including the anatomical terms you need to know. Chapter 3 explores the wide world of beef cattle breeds and types so you can find the best animal for you.

Chapter 1

The Rewards and Responsibilities of Raising Beef Cattle

In This Chapter

- Surveying the benefits of owning beef cattle
- Taking care of your cattle's basic needs
- Creating a budget for your finances and time
- Putting together your advisory team
- Contemplating safety and legal concerns

Raising cattle can be fun, educational, profitable, and downright entertaining. Cattle provide you and your family with wholesome food, and they can put a few extra dollars in your pocket as well. But before you get too wrapped up in all the different opportunities that come with raising cattle, you need to take a moment to look at the investments you need to make in order to properly care for your animals.

This chapter helps you understand what it takes to raise cattle. We show you the positives associated with owning cattle, and we also explain what's involved in meeting their basic needs. Next, we discuss how much money and time you need to raise cattle. Fortunately, you don't have to go it alone when caring for cattle, so we also give you pointers on how to put together a bovine advisory team. Of course, safety should be your paramount concern when raising cattle, so we conclude with some ideas about how to keep yourself and your cattle safe and on the right side of the law.

Burgers and Beyond: Examining the Benefits of Owning Beef Cattle

For more than 30 years, we've had the privilege of caring for cattle. These amazing creatures continually impress us. They have provided so many opportunities and benefits for our family. And they can do the same for yours. Consider some of the main benefits:

- They produce food and income.
- With proper grazing management, they improve the health and productivity of your soil.
- They teach you a lot about animal husbandry.
- They give you the opportunity to develop friendships with other cattle farmers and beef customers.

It's exciting to think about what owning cattle can mean for you and your family! We discuss these benefits in the following sections.

Becoming more self-sufficient

Producing the very food that's served at your kitchen table is a noble goal and one that's shared with an ever-increasing number of people. A single beef animal can yield from 300 to 550 pounds of meat depending on its size. Speaking from personal experience, knowing you have plenty of food in the freezer for dinner every evening is a good feeling. Just as important, raising your own beef cattle gives you peace of mind and a sense of pride by knowing where your food came from and that you raised it yourself.

Can you raise your own beef at a price comparable to what it would sell for in the grocery store? It depends. If you're starting from scratch with your cattle enterprise and need to build fences and shelter and purchase supplies, it may take you some time to recoup those costs. Having some of the basic infrastructure in place reduces your expenses.

If you're a big bargain shopper, producing your own beef costs more than buying everything on sale at the store. However, if you're willing to pay extra for beef raised in a certain manner (without antibiotics or on pasture, for example) and like the higher value cuts of meat (such as steaks or extra-lean ground beef), producing your own beef can be comparable in cost to buying similar quality products at the grocer.

Producing extra income

If you're already raising a beef animal or two for your family and have the space and feed resources, it doesn't take much more effort to raise a few more animals to sell at a profit. You'll find that it doesn't take any additional time to open the gate and rotate eight head of cattle to a new pasture than it does to open the gate for five animals!

Raising additional cattle helps reduce the per-head production cost. Spreading the fixed costs of things — such as taxes, insurance, and (to a certain extent) fences and facilities — over more animals enables you to raise each animal at a lower cost.

In Part IV of this book, we provide more details on ways to make money with cattle.

Improving the land naturally

Well-managed cattle-grazing can improve the soil and diversity of plant species. Research and on-farm experience has shown that controlled grazing by cattle herds

- Increases the number of earthworms and the amount of organic material in the soil
- Leads to a balanced mix of plants
- Increases the production of nutritious edible material per acre

And guess what? A pasture that has been properly grazed is also pleasing to the eye and can help increase the value of your property. For more on working with your land and cattle, head to Chapter 6.

Providing a fun and educational family project

Many families enjoy working together on their beef projects. Caring for cattle has many different facets, so people of all ages and skills can play a role. Because cattle shows occur all over the country, raising cattle also provides your family a chance to exhibit your animals and see the country together. The opportunity to travel with your family and work together can make for some wonderful memories and learning experiences. Check out Chapter 14 for more about showing cattle.

Even if your family doesn't become involved in showing, raising cattle still provides the chance to develop dependability and a good work ethic. Cattle need consistent care everyday — rain or shine. Your family can learn about animal husbandry and develop a greater appreciation for other living creatures. Watching your budget and making decisions about purchases can also help teach important money management skills.

The chapters in Part II delve into the daily chores and animal husbandry routines you take on when raising cattle.

Providing the Basics for Your Cattle

To get off to a good start with your bovine adventure, you need to supply your cattle with a few basic needs: wholesome feed and water, shelter, and a safe environment. Be sure you can fulfill these minimum requirements before you begin or add onto your herd. We explain these requirements in the following sections.

Wholesome feed and water

Your cattle depend on you every day to provide them access to good, clean food and water. The nutrients your cattle need include protein, energy, minerals, and vitamins. The amount of nutrients your animal requires depends on its age, stage of production (growing, reproducing, lactating, or maintaining), performance level, and weight. It also depends on the weather in your region.

A quality pasture can provide most of what your animal needs, but you may have to provide a free-choice vitamin/mineral mix as well. Cattle with high nutritional requirements, such as finishing market animals or cows in late gestation or early lactation, may need more energy or protein than what a pasture can supply. In those cases, you may also need to feed them high-quality hay or some grain.

As a general rule of thumb, figure that your cattle will eat an amount equal to 2.5–3 percent of their body weight in dry feed a day. So a 1,200-pound cow consumes about 30 to 36 pounds of feed daily. If the feedstuff has a high water content like wet, lush pasture, the animal needs to eat even more to get the required level of nutrients.

The water source you provide your cattle can be a natural one like a spring, stream, or pond, or it can be a man-made one like a stock tank or automatic waterer. Whatever you use, just make sure the water is clean and fresh at all times. It needs to be cool and plentiful in hot weather. A nursing cow can drink up to 18 gallons a day. During the winter, don't count on the cows consuming enough snow to account for their water needs. Provide them with a temperate supply of water also.

We get into the specifics of feeding and watering your animals in Chapter 5.

Clean, spacious living and eating areas

Cattle don't need fancy living accommodations, but their facilities should be tidy and not too crowded. They should protect the animals from winter weather and shade them from the sun in the dog days of summer.

- **Winter:** Because they have thick coats, a simple three-sided shed should be sufficient to protect them from wet, winter weather. Weak calves may need additional shelter, however. Plan to provide 75 to 100 square feet of barn per adult animal. Make sure the floor is clean and dry and bedded with straw, sawdust, or woodchips.
- **Summer:** The same type of barn described in the preceding bullet can serve as a shade source during the summer. Well-positioned trees in the pasture are another great shade option. If you plan to confine the animals to an earthen pen along with their barn, provide 250 to 500 square feet of pen space for each animal.

We talk more about setting up shelters for your cattle in Chapter 4.

The amount of pasture your cattle need varies depending on your management, the season, the type of grass you grow, and the amount of precipitation you receive. Pastures that are continually grazed usually can't support as many animals per acre as pastures that are *rotationally grazed* (animals are in a section of the field for a day or two and then moved to another section). Rotational grazing encourages the cattle to eat a wider variety of forages and gives the pasture time to regrow and recover before being eaten again.

During the peak growing season of May and June, a 1⁄2 to 3⁄4 acre of pasture may be sufficient for a 1,000-pound animal. But as grass and legume growth slows in July and August, you may need to provide 2 to 3 acres per animal. During the beginning (April) and ending (October) of the growing season, per-head acreage requirements may increase to anywhere from 5 to 10 acres per head if you don't plan on supplementing your pastures.

Keep in mind that different types of feedstuffs yield varying pounds of product per acre. One acre of stockpiled fescue grass may be enough for a mature cow in early winter, but as its quality declines over time, she may need twice that amount in early spring. You want to select the forages that perform best for your specific soil, weather, and grazing needs. For details on managing your pasture, head over to Chapter 6.

Mud can be one of your biggest challenges when raising cattle. For areas where cattle congregate, such as around water troughs, feeders, or walkways, a heavy use area protection (HUAP) ground covering can make for a cleaner and more pleasant environment. For more on HUAP zones, see Chapter 4.

A low-stress environment

One of the best things you can do for your cattle's health and well-being is to provide them with a calm environment. The good news is that such an environment is good for the herdsman as well! Cattle like predictable, positive routines. Feed and check your cattle at about the same time every day so they know what to expect.

You want your cattle to associate you with good experiences. You can achieve this rapport with your animals by moving them to a new pasture regularly, giving them corn or alfalfa hay treats, or, for very tame animals, giving them a scratch under the chin. Use a quiet voice or remain silent when working with your cattle. Loud noises and quick movements upset them. Be patient when handling cattle as well.

Refer to Chapter 8 for tips on interpreting cattle behavior and handling them properly.

Planning for the Labor and Financial Commitments

Raising cattle can be a rewarding and profitable experience, but you shouldn't take the responsibility involved lightly. Do plenty of thinking to make sure you're ready for the daily care of these living, breathing animals that depend on you for their well-being. You also want to be sure you have the money to maintain your herd, including a financial cushion to handle a large, unexpected vet bill or a big spike in feed prices. The following section highlights some of the obligations you may encounter.

Reviewing daily chore commitments

The amount of time you spend with your cattle for routine care can be as little as a few minutes a day to several hours a day. It just depends on your management style and your situation.

For instance, if your cattle are on pasture or a self-feeder, you may just need to do a quick inventory count, look for any signs of illness or injury, and check that clean, fresh water is available. If, on the other hand, you're preparing cattle for exhibition, you need to take more time to feed them individually and attend to their grooming needs.

Whatever management style best fits your situation is fine, but the key is to regularly observe your animals so any small issues with their health, feed, water, or housing can be fixed before they become big problems.

Planning for big projects and emergencies

To give your cattle the best of care, be sure to make arrangements for large-scale projects. Some tasks, such as annual healthcare work or barn cleaning, need to be on your to-do list every year and may take half a day or more.

Other big projects, such as building or repairing fences, occur less frequently and may involve several days or weeks of work.

Also, if you have commitments other than your cattle (and even the most dedicated herdsman does), have a trusted helper on hand in case of emergency. This person can care for your cattle if you're unavailable.

Similarly, exchange cellphone numbers with your neighbors, and make sure they have yours so you can be in contact if problems arise. You may even find it helpful to post signs on your fences and barns with your contact numbers so if someone runs into your fence and leaves a big gaping hole they can reach you. (This advice may sound silly, but we've had this exact situation happen to us on more than one occasion, so take heed, my friend!)

Looking at purchase prices

As is the case with most anything you buy, you can pay a little or a lot for cattle, and most often you get what you pay for. Here are some of the options available to you:

- **Bottle calves:** One of the least expensive ways to get into the cattle business is with a bottle calf. However, keep in mind that you'll have a bigger investment in terms of your time commitment in feeding and caring for the youngster. Bottle calves usually sell on a per-head basis and can be anywhere from $50 to $200.
- **Older, weaned calves or yearlings:** These animals sell by the pound, and the price varies greatly depending on the time of year, the quality and weight of the cattle, and the overall demand. In the last ten years, prices for healthy animals have ranged from $0.80 a pound up to $1.60 per pound.
- **Breeding and show stock:** These bovines sell by the animal. A young show heifer or steer may cost a bit more than a commodity market animal or may go up in cost to several thousands of dollars if you want an animal that has a chance of being a state fair grand champion. You can find a quality bred, middle-aged cow for around $1,100 give or take several hundred dollars, depending on market conditions.

Check your local agricultural publications or visit a local auction to get a feel for fair prices in your area.

Calculating feed expenses

After the initial purchase price, feed is your biggest variable expense in raising cattle. So you need to set aside enough funds to be able to properly feed all of your cattle. You may have feed expenses associated with grain rations and forages like hay or pasture as well as minerals, vitamins, and supplements.

The cost can vary greatly depending on your resources, the type of cattle you're raising, and the weather. Traditionally, allowing cattle to graze is less expensive than feeding hay. Higher-producing animals like nursing cows or finishing market animals have higher feed expenses. Usually the feed bill for winter cattle feeding is the biggest because the animals are consuming more harvested feeds. But in spring and summer, your direct costs may be limited to the cost of pasture plus minerals and vitamins.

If you're growing yearlings out on your excess pasture, realize that you have an opportunity cost associated with the pasture: You could be getting paid rent by another cattle producer for the use of that pasture. *Custom grazers* are cattle producers who care for cattle belonging to someone else. They charge anywhere from $0.50 to $1.50 per day to graze yearlings or cows, so for a true accounting of your feed costs, you need to "charge" your grazing animals a comparable rate when figuring your profitability.

Since the turn of the century, the feed cost for an animal eating a grain-based ration to gain 100 pounds has been as low as $40 and as high as $100. The overall trend has been for the cost of feeding cattle to continue to climb higher.

Your best course of action to determine your feed costs is to keep excellent records of your feed expenses and animal production so you can figure how much money it takes for your particular situation. Until you have an opportunity to collect the data, estimate your feed costs on the high side so you don't come up short financially.

Figuring the cost of healthcare

Healthcare falls into two main categories: preventive care and emergency care. Depending on the routine needs of your herd, you can spend $7 to $20 per animal on vaccinations and parasite treatments every year whether you do it yourself or have the vet do it for you. If you hire the vet to do any castration, dehorning, or pregnancy checking, add another $5 to $15 per animal.

Expenses associated with emergencies such as illness, injury, or pregnancy complications can mount up quickly. Some of the older antibiotic treatments for infections cost less than a dollar per animal but newer medicines can be $10 to $20 to treat one animal. It's also important to plan for emergency vet charges.

Most vets have a flat fee for visiting your farm, and then they charge an hourly rate in addition to that fee. Inquire about the amount your vet charges for services so you can allocate funds for unexpected healthcare needs.

Budgeting for facilities and land

The main housing costs are for the pasture, shelter, and fencing, and they can vary tremendously depending on what you already have and what you're looking to have.

If, for example, you already have a few fenced acres with a shelter, your start-up costs for facilities and land will be nil. For brand new endeavors, on the other hand, your pasture rental could range from $10 to $60 an acre. Of course, if you have the time and skills, you may be able to trade some sweat labor for access to pasture and improve a neglected piece of land into a picturesque bovine smorgasbord.

After you have your land, you then have to figure out what types of facilities and fencing you want for your cattle. You can build a basic three-sided barn for around $400 to $600, or you can buy a small, prefabricated structure for around $1,000. Installing barbed-wire, high-tensile, or woven-wire fencing costs anywhere from $1 to $3.50 a foot depending on the terrain, the type of fence you need, and whether you do the work yourself or hire someone else to do it.

Chapter 4 has more information on shelter and fencing options.

Developing a Herd Mentality: Assembling Your Team of Experts

When it comes to giving your cattle the best of care, be proactive about networking and learning from others. Even the most experienced cattle farmer hasn't encountered every possible bovine problem or had the chance to use all the newest products. Consulting with a mentor, vet, extension agent, or feed salesperson can greatly contribute to your knowledge base and help make you a better herdsman or herdswoman. We clue you in on your possible support system in the following sections.

Making friends with a mentor

A mentor can help you learn more about raising cattle without going through the stress of trial and error. He or she can help you network with other producers and potential customers. Find someone who meshes well with your personality and has a similar way of doing business.

You can look for a mentor in the area where you live or through a beef cattle or livestock trade association. More and more cattle producers are becoming active online, so you may even be able to build up an electronic relationship with someone who's not even in your town or state!

Vetting your veterinarian

A good working relationship with your vet can be crucial to the success of your cattle-raising experience. A vet can guide you in developing a preventive health program for your herd, help figure out ways to solve problems with your cattle's health and well-being, and be a literal life-saver in emergency situations.

Qualities to look for in a good vet include the following:

- **Experience with and interest in working with cattle:** Many vets specialize in caring for pets or horses. Your best bet is to have a vet whose main focus is cattle, because some things, such as pregnancy checking or abnormal birthing presentations, become easier to handle with lots and lots of practice. A vet with interest in cattle is more likely to remain up-to-date on all the latest treatments and issues in cattle production.
- **Easy to communicate with and willing to answer questions:** If you are new to the cattle business or are interested in giving your cattle the best of care, you will have questions. You should be comfortable talking with your vet about these questions, and she should be able to discuss the issues with you in a patient, professional manner.
- **Accessibility:** Does the vet come to your farm, or do you have to take your cattle to the clinic for care? Can you schedule appointments for routine care in a timely fashion? In case of emergency, can you reach your vet easily and get help within a reasonable amount of time? These are all good questions to explore with any vet you are considering. You can even ask the vet for some references so you can get answers to these questions.

Relying on local extension agents

Extension agents are people affiliated with land-grant universities. They're available to provide useful, research-based information to agricultural producers and others. Most counties have a cooperative extension office; you can also find regional and state personnel. To find contact information for your local extension agent, go to `www.csrees.usda.gov/Extension/USA-text.html`.

Your county extension staff may or may not be experts in beef cattle production, but they almost always can help you find someone to answer your questions about raising cattle. Usually the state level of the extension service has experts in the areas of cattle nutrition and health, pasture management, livestock evaluation, and agricultural/food rules and regulations. Most extension information is available at little to no cost.

Asking for advice from your feed dealer

If you aren't pasture feeding or need to supplement with a commercial ration, you have some thinking to do. Deciding on an economical, balanced ration for your cattle isn't always an easy task. Luckily, feed dealers can help you sort through the options and develop a feeding program to fit the needs of your animals. Do an Internet search to find the national or regional feed companies and their local dealerships.

Holy Cow! Important Safety and Legal Considerations

To properly care for the animals and the people involved in your cattle-raising endeavor, you need to be aware of potential safety and health issues. You also need to plan how you can be a responsible cattle owner and member of your community.

Preventing physical injuries

For the comfort and well-being of yourself, your cattle, and your helpers, you must handle your animals correctly and provide a safe environment for all involved. Here are some ways to prevent human and bovine injuries:

- **Don't take the good behavior of your cattle for granted or become complacent when handling them.** All animals — including tame, domesticated cattle — are unpredictable, so always be on guard around them. Chapter 8 provides tips on recognizing cattle body language.
- **Use extra caution around bulls, cows with calves, or animals that are handled infrequently.** Because these animals can be quite territorial or unaccustomed to humans, they may act aggressively or in an unpredictable manner.

- **Realize that cattle perceive their world differently than humans.** They have good panoramic vision but can't see directly behind them, so they get spooked if you approach from the rear. Because of their limited vision, they also have poor depth perception and balk at shadows. They're sensitive to high-pitched noises or unexpected sounds as well. For more tips on handling cattle, see Chapter 8.
- **Avoid handling cattle during very hot or cold weather or if the footing conditions are slick due to ice or mud.** Extreme weather conditions put stress on cattle, so you don't want to increase the chance of illness or injury by handling cattle in these less-than-ideal conditions unless absolutely necessary. Also, be sure to use nonslip flooring in all barns, handling areas, and walkways.
- **Construct sturdy handling facilities that capitalize on cattle's natural behaviors.** By designing facilities that are in tune with their instincts, you put less stress on the animals when handling them. Also, leave yourself an escape route when working around cattle. Chapter 4 has numerous ideas on handling facilities.
- **Keep trash and loose items picked up.** Debris tossed by the wind can cause cattle to spook. Pieces of garbage also can be lodged in the hooves or accidentally ingested.
- **Maintain your fences.** Sturdy fences help keep cattle safely contained and off the road and out of crop fields. They also reduce the chance of injury due to cuts and scrapes from loose wire.

Watching out for zoonotic diseases

Zoonotic diseases are diseases that can be transmitted between humans and animals. They can be spread through the air, by direct contact, by touching a contaminated object, through oral ingestion, or by insect transmission.

Some of the zoonotic diseases of concern in the United States include brucellosis, campylobacteriosis, leptospirosis, ringworm, and tuberculosis. For a detailed description of each of these diseases, see Chapter 10.

To help prevent zoonotic diseases

- Wash hands with soap and hot water after working with animals and before eating or drinking.
- Wear disposable gloves when handling animals suspected to have disease.
- Avoid unpasteurized milk and milk products.
- Cook meat thoroughly.
- Stress the importance of good hand washing and proper hygiene (no thumb sucking, please) to all children who spend time around cattle.

Pregnant women shouldn't assist with birthing or handling sick animals to avoid harming the pregnancy.

Considering your insurance needs

The three types of insurance you should consider when raising cattle include the following:

- **Personal and farm liability insurance:** This coverage provides payment for your legal liability for damages due to bodily injury or property damage. For example, farm liability insurance may protect you if your farm causes pollution of nearby properties due to pesticide or animal manure runoff. Additionally, if your cattle escape from their enclosures and injure another person's body or property, this type of liability insurance may provide coverage.
- **Farm personal property and building insurance:** This insurance can be used to protect barns, farm equipment, and livestock. Be sure your coverage provides compensation if your animals are accidentally shot, drowned, attacked by wild animals, or electrocuted.
- **Liability insurance:** If you're selling beef for human consumption, get insurance coverage for product liability issues. This coverage helps protect you against damages due to food quality and safety.

Being a good neighbor

Even though you're excited about having cattle in your backyard, your neighbors may not share your enthusiasm. To avoid awkward and tense confrontations with the folks who live near you, take these steps to foster good neighborly relations:

- Obtain any needed permits or zoning changes before you build fences or barns or bring cattle to your property.
- Display any certifications you received that indicate you have taken steps to be a good environmental steward, including recognition for soil and water conservation efforts or certification in humane animal care and handling.
- Reach out to your neighbors to keep them informed of any activities you may need to do that temporarily cause an increased amount of dust, noise, or smell.
- Have an open house so your neighbors can see all the positive animal husbandry practices you use.

- Situate potential problem areas like animal corrals, feeding areas, or compost piles as far away and downwind from the neighbors as possible.
- Keep your barns, fences, animals, and equipment clean and tidy to enhance the public impression of cattle farming.

Complying with zoning regulations

Just because you want to raise cattle on your 10-acre ranchette on the edge of town doesn't mean you can. Where you farm depends on the zoning laws for your community. Before you even consider starting a cattle farm, visit your local zoning department to find out whether your property is zoned for agricultural use. If it is, find out about any specific laws regarding livestock, such as the number of animals you may have or the amount of land required.

If your land is zoned for some other use besides agriculture, you probably need to apply for a variance. When you apply for this variance, the zoning office personnel contacts your neighbors to notify them of the requested change and then publishes a notice in the local paper. You also have to present your case to the zoning board. It will either approve or deny your request. If you're rejected, you can file an appeal.

Chapter 2

Bovine Basics: Understanding Cattle Terminology and Anatomy

In This Chapter

- Using the correct terminology for cattle
- Identifying bovine body parts and reproductive anatomy
- Understanding important differences between beef and dairy cattle

Often when people think of cattle, the image of a big, white cow with black spots and a huge udder comes to mind. This basic idea of cattle isn't wrong, but if you want to raise beef cattle, you need to expand your knowledge a bit. This chapter aims to help you better understand what you read and hear about cattle.

We start by introducing the various ways that cattle farmers refer to their animals based on their age, sex, and stage of life. Because you also want to be able to knowledgeably describe cattle's anatomical features when choosing your animals and seeking feeding, breeding, or healthcare advice, we identify the names and functions of their various body parts. This rundown of information allows you to effectively communicate about your cattle. Finally, this chapter touches on some of the differences between cattle bred for milk production and cattle bred to provide meat.

Livestock Lingo: Describing Cattle by Age, Sex, and Life Stage

When is a cow not a cow? Although this question may seem like a silly riddle, people often use the label "cow" incorrectly to refer to cattle of any age or sex.

The term *cow* actually refers to a female bovine that has given birth and is 2 years or older. To ensure that you buy and sell the right sex and age of cattle, you need to be able to describe them correctly. Don't worry. In this section, we steer in you in the right direction.

Take a few minutes to get acquainted with the terms in Table 2-1. They accurately describe various types of cattle. Note that some of the terms depend on both the animal's age and sex (heifer, dam, steer, bull) and that others are used regardless of gender and are more a description of life stage (calf, weanling, fat cattle).

Table 2-1 Cattle Names

Cattle Name	*Description*
Bred cow	A pregnant cow
Bred heifer	A pregnant heifer
Breeding stock	Bulls and heifers/cows whose main purpose is to be sires or dams
Bull	A male of any age with an intact reproductive system
Bull calf	An intact male calf
Calf	A young (7 months or younger) bovine of either sex
Cow	A female bovine that has given birth and is 2 years or older
Cull	A cow or bull that's no longer productive (A cow may no longer be able to become pregnant or produce enough milk to feed a calf. A bull may be infertile or no longer physically capable of mating.)
Dam	The mother of an individual bovine
Fat cattle (fats)	Cattle that have reached slaughter age (13–30 months) and weight (1,000–1,400 pounds) (These cattle, which are almost always steers or heifers, aren't fat to the point of obesity, but they have reached the point in their lives where muscle and bone growth has slowed or stopped and fat deposition is occurring.)
Feeder cattle	Cattle that are weaned and ready to or have started eating a diet high in energy and protein
First-calf heifer	A female that's either pregnant with or is raising her first calf
Heifer	A female bovine that hasn't given birth and is usually 30 months of age or younger
Heifer calf	A female calf
Herd	A group of cattle
Open cow or heifer	A female that is not pregnant
Pasture exposed	Heifers or cows that have been living in the same pasture as a bull and are assumed to be pregnant (The pregnancy and exact birthing date haven't been confirmed by a vet's examination.)
Sire	The father of an individual bovine

Cattle Name	Description
Steer	A castrated male
Stocker cattle	Weaned cattle that are on a forage-only diet up until 10–12 months of age
Weanling	A bovine of either sex that has recently stopped nursing (usually 5 months or older)
Yearling	A bull, steer, or heifer that's around 12 months of age

Some cattle names and descriptions vary by region of the country, so always clarify the history and physical condition of the cattle before you buy. For example, sometimes bulls are sold as feeder calves. However, if you plan to grow these calves out and sell them as fat cattle, you should be aware that the price for fat bulls is greatly discounted compared to fat steers or heifers (because the meat from bulls is often of lower eating quality compared to steers or heifers). You can always castrate feeder calves that are intact males, but the procedure involves a slight risk of infection.

Grasping the Wonders of the Bovine Body

All types of cattle have some unique body parts that you need to be familiar with so that you sound like you know what you're talking about when discussing your cattle. We start with the general anatomy, including the bones, the muscles, and other structural landmarks.

Then we move on to the stomach, one anatomical feature of cattle that's particularly fascinating, not to mention important for bovine health. Later in this section, we take you on a tour of this multipart organ that enables cattle to digest foods that many other animals can't eat.

And if you're interested in breeding cattle or want to be sure that your females don't become pregnant, you need a solid understanding of the male and female reproductive systems. We provide illustrations of these anatomical features and offer insights into their functions.

"Nice pin bones": Getting acquainted with general cattle anatomy

If you want to accurately discuss your cattle questions and concerns, you need to be familiar with the names and functions of their various anatomical features. Figure 2-1 identifies key cattle body parts.

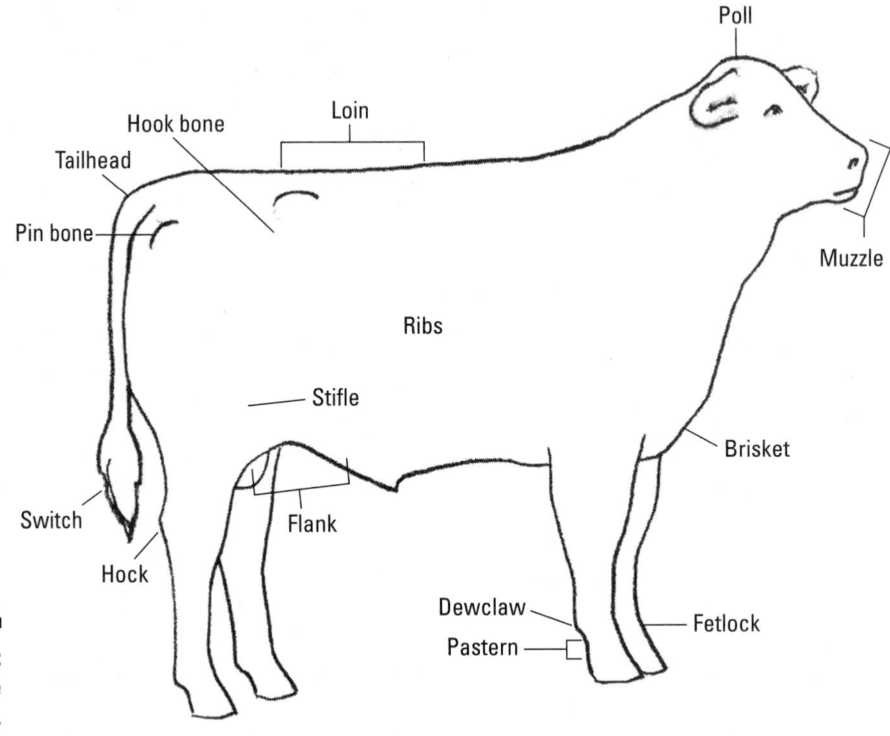

Figure 2-1: Beef cattle anatomy.

The body parts we describe in this section are found on both males and females; we discuss the specific male and female reproductive anatomy in later sections.

- **Brisket:** The fleshy area of muscle and fat between the beef animal's front legs.
- **Dewclaw:** A small cartilage-like protrusion located above the pastern on all four legs.
- **Fetlock:** Joint above the hoof in the area of the dewclaw.
- **Flank:** Area in front of the hind leg that ties into the underside of the belly region.
- **Hock:** The joint on the back side of the hind leg.
- **Hook bone:** Part of the pelvic bone that juts out below the loin and behind the ribs.
- **Loin:** Region of the back where many high-quality steaks are located. Cattle with long, full loins should yield more dollars' worth of meat.
- **Muzzle:** Area of the head made up of the nose and mouth.
- **Pastern:** The region between the dewclaw and hoof.

- **Pin bone:** Part of the pelvic bone that sticks out on either side of the tail.
- **Poll:** The top of the head (it may be pointy or block-shaped depending on the animal), which sprouts a long lock of hair.
- **Ribs:** The bones coming off the backbone and down over the abdominal cavity.

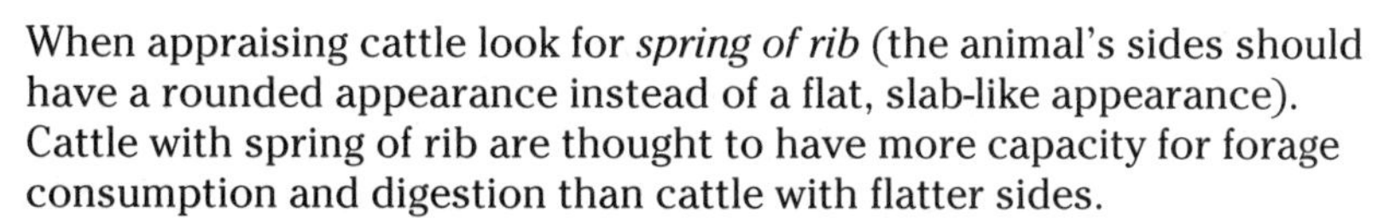
When appraising cattle look for *spring of rib* (the animal's sides should have a rounded appearance instead of a flat, slab-like appearance). Cattle with spring of rib are thought to have more capacity for forage consumption and digestion than cattle with flatter sides.

- **Stifle:** The joint directly above and behind the flank.

Bulls and cows can injure the stifle when mounting other cattle. Cattle with stifle injuries are lame and may have an abnormal bulge in the stifle area. This injury is difficult to treat, and most afflicted cattle are butchered.

- **Switch:** Part of the tail with long hair.
- **Tailhead:** Area at the end of the bovine where the base of the tail connects to the body. For the sake of appearances, the tailhead should lay smoothly into the back and not stick upward.

Ruminating on cattle stomachs

Beef cattle can thrive by grazing land that's too arid or fragile for growing food — grains, vegetables, or fruits — for human consumption. Cattle have this dietary edge because they're capable of consuming plants that people can't eat. The cattle convert the plants to nutrient-dense beef that humans can consume. The key to cattle being four-legged food factories is their stomach. To truly appreciate, care for, and harness the potential of your cattle, you need to understand this fascinating organ.

Cattle — along with about 150 different domestic and wild species, such as sheep, goats, deer, and giraffe — are ruminants. *Ruminants* have four-part stomachs, and they regurgitate their partially digested food and rechew it in a process known as *cud chewing.*

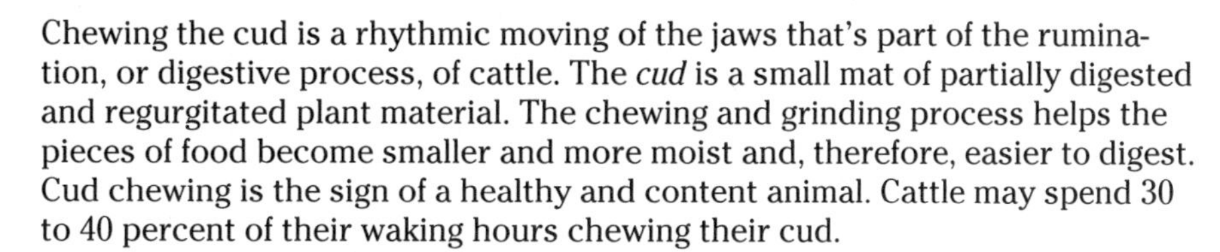
Chewing the cud is a rhythmic moving of the jaws that's part of the rumination, or digestive process, of cattle. The *cud* is a small mat of partially digested and regurgitated plant material. The chewing and grinding process helps the pieces of food become smaller and more moist and, therefore, easier to digest. Cud chewing is the sign of a healthy and content animal. Cattle may spend 30 to 40 percent of their waking hours chewing their cud.

Here are the four parts of the cattle stomach listed in the order by which the food moves through them (Figure 2-2 shows these parts):

- **Rumen:** This part of the stomach serves as a storage space and, in an adult bovine, can hold more than 20 gallons of food. After the food is eaten, it's stored here while it awaits cud chewing. The rumen also serves as a large fermentation vat. The bacteria and protozoa residing here break down high-fiber feeds, such as grass, hay, and corn stalks, that nonruminants can't digest.

The fermentation process produces lots of gas that ruminants eliminate by belching. If anything impedes this belching, gas builds up in the rumen and causes a dangerous condition called *bloat.* To find out what you can do to prevent and treat bloat, check out Chapter 10.

- **Reticulum:** This pouch-like structure lies in the front of the abdominal cavity. Stomach contents move freely back and forth from the reticulum and rumen, so the reticulum can aid in the rumination process. Heavy feed or foreign objects fall in this compartment.

If cattle accidentally eat nails or other sharp items, the items can become lodged in the reticulum walls causing hardware disease. Your vet may have the afflicted animal ingest a magnet to collect the metal and keep it from damaging the reticulum.

- **Omasum:** This compartment is made of many layers of tissue, like pages in a book, and primarily serves to absorb water and nutrients into the body.
- **Abomasum:** Also called the true stomach, the abomasum secretes digestive juices and enzymes that further break down feed.

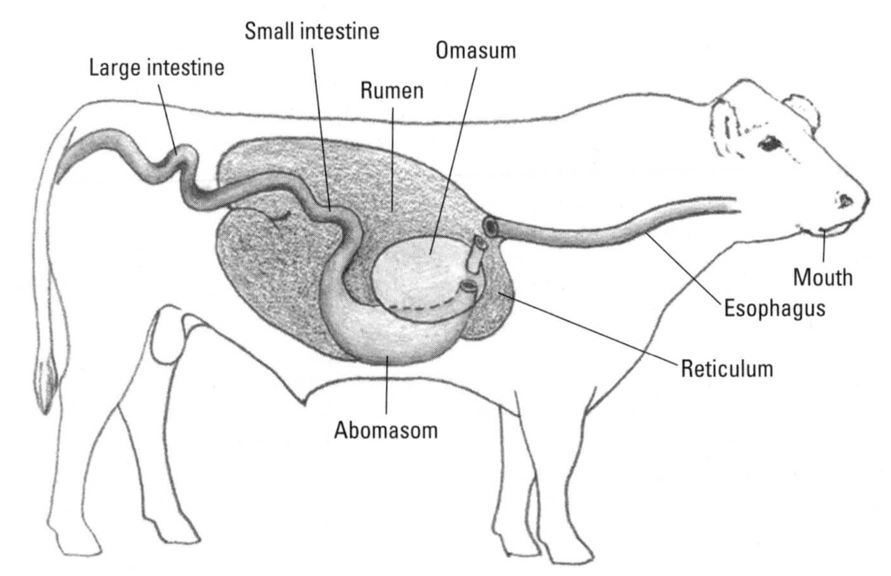

Figure 2-2: The stomach and digestive tract of a beef animal.

Checking out male reproductive anatomy

On a cattle farm, a bull has one purpose: to breed. To achieve his goal, he needs a healthy and functioning reproductive system. By knowing the name and functions of the various reproductive parts, you can be on the lookout for problems.

Losing an entire year's worth of calf production because your bull is infertile can be a hard financial blow to overcome. So before purchasing a bull, make sure he's capable of breeding cows. You can find out by confirming that he has recently mated with and successfully bred cows or by having the bull undergo a breeding soundness exam (refer to Chapter 11 for details).

A bull's reproductive tract (see Figure 2-3) comprises the following:

- **Testicles:** The testicles not only produce the *spermatozoa,* the male reproductive cells, but also the male hormone, *testosterone.* Testosterone is needed for the maintenance of the male reproductive tract and for normal sex drive. The testicles are contained outside the main body cavity in the scrotum.

  The location of the testicles is critical because normal sperm formation occurs at a temperature several degrees lower than the body temperature of 100 to 103 degrees. Sperm that have been exposed to high heat either through weather extremes or a malfunctioning scrotum aren't capable of fertilization.

- **The secondary sex organs:** The secondary sex organs, which include the epididymis, vas deferens, and penis, transport the spermatozoa from the testicles to the female reproductive tract (see the next section). Spermatozoa are mixed with liquids secreted by the seminal vesicles and prostate gland (which are accessory sex glands) to form semen. Blockage or infection of any of these parts could render a bull infertile. Bulls can be castrated by clamping and crushing the spermatic cord (for more detail on castration techniques, refer to Chapter 12.)

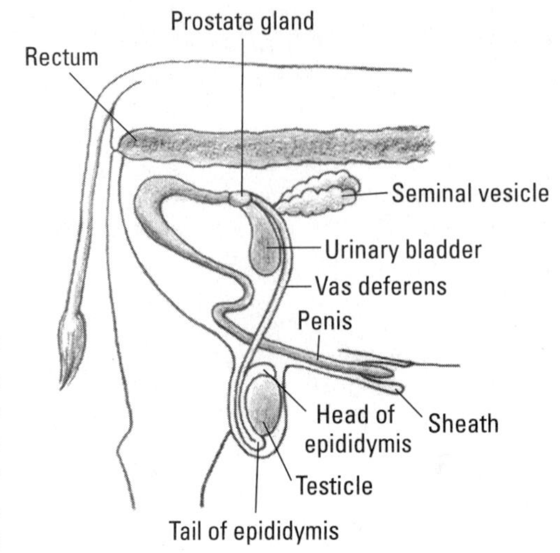

Figure 2-3: The reproductive anatomy of a bull.

Making sense of female reproductive anatomy

Along with the fertility of its bulls (described in the preceding section), one of the keys to profitability on many farms is the reproductive ability of its heifers and cows. If the cows don't become pregnant, you won't have a calf to sell or beef to eat. To keep your cows healthy and productive, become familiar with the inner workings of their reproductive systems (see Figure 2-4).

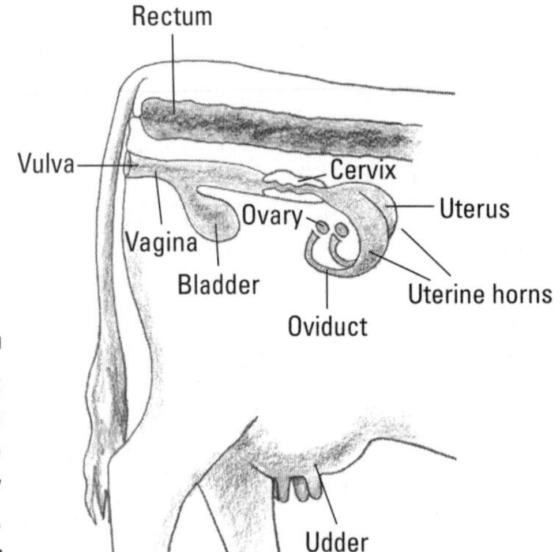

Figure 2-4: The reproductive anatomy of a cow.

The main parts of the female reproductive tract include the following:

- **Cervix:** Semen must pass through this reproductive organ to reach the egg. During pregnancy, the cervix becomes filled with a mucous plug that helps protect the uterus and developing fetus from infectious organisms. During calving, the cervix relaxes and expands so the calf can pass through.
- **Ovary:** This is the main reproductive organ of female cattle. These small organs — each cow has two of them — produce the female reproductive cell (egg) plus the estrogen and progesterone hormones.
- **Oviduct:** When the egg is released from the ovary, it moves to the oviduct, where fertilization occurs.
- **Udder:** Although it is not a reproductive body part, only cows have udders. This mammary tissue is located between the hind legs and produces milk.
- **Uterus and uterine horns**. One of the cow's two uterine horns receives the fertilized egg from the oviduct. The embryo grows in the uterine horn because the uterus is small and not well-developed. In cattle, the uterus basically serves as a passageway connecting the uterine horn and the cervix.
- **Vagina:** Semen from the bull is deposited into this canal, and the calf passes through it at birth.
- **Vulva:** The external opening of the vagina through which both the calf and urine are passed.

Meat or Milk? Investigating Different Types of Cattle

Over the years, farmers have bred cattle to optimize certain traits. They have given particular attention to two traits: cattle's ability to provide meat and their ability to produce milk. This breeding has lead to the development of two distinct cattle types:

- **Beef cattle:** These cattle excel at efficient growth of muscle and, hence, beef production. Head to Chapter 3 for a rundown of the different breeds of beef cattle.
- **Dairy cattle:** These cows are known for their tremendous milk-producing abilities.

In the following sections, we provide an overview of the major differences between dairy and beef cattle. Of course, we assume that, because you're reading this book, you're interested in raising beef cattle, but knowing a bit about dairy cattle helps.

Looking at the udderly fascinating world of dairy cattle

You can't help but admire dairy cows. Holsteins, the most popular dairy breed, are often statuesque with striking black and white color patterns and huge udders.

Even more impressive than a dairy cow's looks is her ability to produce milk. A productive dairy cow can yield around 20,000 pounds of milk a year — that's 12 to 14 times her body weight every year! Because dairy cattle devote so much energy to producing milk, dairy animals (both male and female) yield less meat than similar-sized beef animals.

The fate of male dairy calves on dairy farms

Other than a few select calves retained for breeding stock, most male dairy calves born on dairy farms are used to produce veal. *Veal* is meat that comes from calves that are usually 18 to 20 weeks old and weigh from 475 to 500 pounds. The male calves are fed a milk-based diet to help the meat retain its light color and mild flavor.

If you live in or near a large dairy producing state like California, Wisconsin, New York, Idaho, or Pennsylvania, you may be able to purchase bull calves for a low price. But if you aren't interested in veal production, what do you do with the male dairy calves? Even though they weren't bred to be meat animals, you can still raise these cattle like you would beef cattle to produce quite nice, tender beef. Just be sure the price is low enough to compensate for their decreased ability to convert feed into muscle. Also, keep in mind that bulls that aren't used for breeding purposes have a reputation for having mean and aggressive dispositions, so you should castrate them (see Chapter 12 for details) to make them safer and easier to handle. Remember, too, that dairy steers command a lower price than beef steers because they have less meat.

The big beef: Eyeing the traits of beef cattle

Cattle farmers have selected beef cattle to convert feed into meat as efficiently as possible. After beef cattle have consumed the minimum nutrient requirements for basic bodily functions, they put on weight mainly in the form of muscle (as opposed to bone or fat). This genetic predisposition to grow lean muscle is why a 90-pound newborn calf can reach butchering size in as little as 13 months.

A beef cow produces an adequate, but not excessive, amount of milk for her calf, which enables her to maintain her physical condition and rear a calf with moderate amounts of feed. (For more on feeding your cattle, flip to Chapter 5.)

In addition to their ability to grow efficiently, the majority of beef cattle are selected to perform adequately in a variety of other important traits, such as reproductive ability, hardiness, and efficient use of feed. These traits are what make it possible for most beef cattle to reproduce and thrive with minimal basic care. In fact, many beef cows can annually raise a calf well into their teens.

Quality beef cattle that yield a good amount of meat for their size have a certain physical conformation. They have a rectangular body shape with a broad chest and width through the shoulders. They have thickness down the tops of their backs, and their ribs and stomachs are full and round, not flat or shrunken. When viewed from behind, the animal appears plump, not skinny or bony. When shopping for cattle, keep in mind that bovines with this “look” are worth more money and put more beef in your freezer. For more on judging conformation, go to Chapter 7.

Beef cattle range in colors from solid white to red, gray, and black. Some sport white blazes on their faces or the tops of their necks. Beef cattle are sometimes spotted, but spots are becoming less common because buyers nowadays pay a premium for solid black cattle (see Chapter 3 to find out why).

Chapter 3

From Angus to Zebu: Looking at Beef Cattle Breeds and Traits

In This Chapter

- Determining which beef cattle traits match your goals
- Checking out the characteristics of British, European, composite, and heritage breeds

One of the first steps to having an enjoyable and profitable cattle raising experience is selecting the breed of beef cattle that is best for you. A *breed* is a group of animals with common ancestors. Animals within a breed share distinctive characteristics that they pass to their offspring. Choosing a breed that aligns with your goals makes the job easier and more productive.

In this chapter, we give you pointers for identifying the cattle traits that fit with your specific goals and environment. We also discuss the differences between purebred and crossbred cattle and the pros and cons of each. We wrap up the chapter with an overview of common beef cattle breeds. By becoming familiar with these breeds' characteristics and qualities, you're better qualified to pick the breeds that best suit your needs.

Identifying the Traits That Are Important to You

The breed of beef cattle you select depends on your cattle raising goals. For instance, do you want cattle to eat a pasture that you're tired of mowing? Or do you want to raise and sell feeder calves? Maybe your family wants to show cattle at the local agricultural fair or raise beef to eat or even sell.

If you're still pondering your cattle-raising aspirations, flip to Chapter 1 to find out more about the benefits of owning beef cattle. You can also turn to Part IV of this book to get ideas about different ways to make money in the cattle business. After you've decided on your reasons for owning cattle, you're in a better position to know what attributes your cattle need to possess so you can be successful and enjoy yourself.

When deciding on a breed, you need to consider a variety of factors:

- Local availability
- Hair color
- The presence (or lack) of horns
- Your climate and feed resources
- Your preference for purebred versus crossbred bloodlines

We explore each of these factors in detail in the following sections.

Keeping up with the Jones's: Follow your neighbors' lead when choosing breeds

Unless you have your heart set on a particular breed because that's the breed you fondly recall your grandfather raising when you were a child or because your customer demands a specific breed, a good starting place for choosing your breed is in your own backyard. Take a look at local cattle farms and note what breeds they have. As we discuss in detail later in this chapter, different breeds of cattle are better adapted to certain types of environments, feeds, and marketing options.

You'll reap the following benefits when you buy breeds of cattle from nearby sources:

- **Bigger selection:** By picking a breed that's common in your region, you'll likely have a bigger selection to choose from when buying cattle.
- **Easier adaptability:** Buying cattle reared close to your home means your new purchases are less likely to have to adapt to new feedstuffs and climate. Cattle in different parts of the country eat different diets and are accustomed to handling the weather particular to their region.

- **Lower transportation costs:** Buying beef cattle close to home means that you save on transportation costs, which can add substantially to the purchase price depending on distance and the number of cattle you're hauling.
- **Proven record of success:** If other local cattle producers are having success with particular breeds, those animals are likely the breed of cattle that thrive where you live and that are easiest to sell.

Purple cows need not apply: Picking your cattle's hair color

Cattle hair comes in about every color of the rainbow (except green and purple), plus it comes in other colors such as brown, gray, and black. The animal's coat can be one solid color, or it can be mottled with several different colors. Cattle may have big spots or patches of white, or little white spots flecked through a dark hair coat (the latter are often described as *roan markings*). Cattle sometimes have white markings on their heads, necks, legs, bellies, or tails.

A trend in cattle hair color has been to prefer solids and, especially, solid black because these cattle sell for more money than spotted cattle. Solid black is the most popular color because it's the best color for hiding flaws like being too skinny or lacking muscle. Black's ability to naturally camouflage flaws may also lead to a higher placing at cattle shows. Just as fashion magazines recommend dressing in black clothes to give you a sleek and flattering silhouette, the same principle applies to black cattle hair!

Additionally, even though it may not always be true, buyers sometimes think that if cattle look the same, they'll grow and eat the same quantity of feed and reach market weight at the same time. Achieving a uniform look for a group of cattle is easiest if they all have solid black hair. Feeder calves that are solid black often sell for 5 to 10 cents more per pound than cattle of any other color.

Black cattle do have one potential drawback: They can have a harder time staying cool than their lighter-colored counterparts. Be prepared to provide extra shade and cool water for black cattle. (Check out Chapters 4 and 6 for more information on shade and watering.)

Horning in: Choosing animals with or without horns

A *horn* is a pointed projection coming from the head of an animal. It's made of fibrous, structural proteins covering a core of living bone. In cattle, the horns come out the side of the head above the ears. Contrary to popular belief, both male and female cattle can have horns.

Unless you're raising a breed of cattle, such as Texas Longhorn or Horned Hereford, in which horns are a standard and expected breed characteristic, it's better to raise cattle without horns. Horned cattle are more dangerous to other cattle and humans, and they also need more space at the feed bunk and during transportation to prevent injury. These increased management challenges mean horned cattle sell for a lower price than cattle without horns.

You need to understand a few terms related to the horn issue so you always know what you're buying. Consider the following:

- **Dehorned:** Describes an animal that has had its horns physically removed. A dehorned animal can still pass on the genes for horns to its offspring. (We discuss dehorning techniques in Chapter 12.)
- **Horned:** Describes an animal with intact horns.
- **Polled:** Describes an animal born without horns.
- **Scurs:** Scab-like growths that form in the same spot where horns grow. They aren't attached to the skull like horns and rarely grow big enough to cause problems.

The genetics behind cattle horns

Being polled or horned is an inherited trait. All inherited traits are determined by the animal's *genes,* which are hereditary units consisting of sequences of DNA. For beef cattle of *Bos taurus* ancestry, the trait of being horned or polled is controlled by a single pair of genes. One gene in the pair comes from the sire and the other from the dam. The gene for the polled trait (P) is dominant to the horned gene (p), so if an animal has one gene for polled and one for horned (Pp), the animal will be polled.

An animal can be further described as *homozygous polled* (PP), meaning that it has two polled genes, or *heterozygous polled* (Pp), meaning that it has one horned gene and one polled gene. A blood test can determine whether an animal is Pp or PP. Because the polled trait is dominant, a homozygous polled animal will have polled offspring even if the other parent is horned.

Cattle with *Bos indicus* ancestors have two sets of genes controlling the horned/polled trait. In addition to the polled (P) and horned gene (p), they also have the African horn gene (Af). A cow with two Af genes or a bull with one or two Af genes will be horned even if it is PP or Pp.

Some breeds of cattle, particularly Angus, Polled Shorthorn, or Polled Hereford, are always polled. In most other breeds, the occurrence of horns is possible but avoidable through selective breeding.

Physically dehorning cattle is stressful to the animal and can lead to complications. Take the time to remove horns through breeding instead. Genetically eliminating the presence of horns is much easier on you and your cattle.

Matching breed characteristics to your feed resources and customer needs

The key to raising healthy cattle and making a profit is balancing your breed's characteristics — such as body size, milk production, and growth rate — with your climate and feed resources and your customer demands. Understanding how all these traits fit together and how they're impacted by the environment the cattle live in helps you pick the best breed of cattle for your situation.

Here are the key breed characteristics you need to consider:

- **Carcass merit:** The *carcass* of an animal is the muscle, fat, and bone left after the head, hide, and most of the internal organs have been removed. *Carcass merit* is an evaluation of the yield, or lean meat, produced by a carcass and the eating quality of the meat. If you sell your cattle *on the grid,* meaning that the price you receive is based on the yield and quality of the carcass (rather than strictly on the live weight of the animal), high carcass merit directly results in higher prices for your cattle.

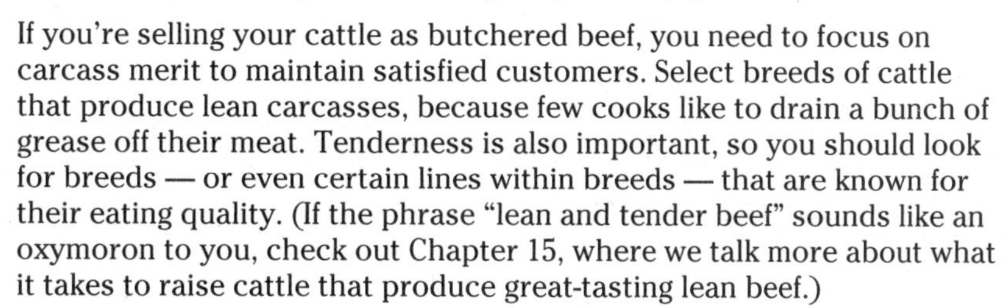

  If you're selling your cattle as butchered beef, you need to focus on carcass merit to maintain satisfied customers. Select breeds of cattle that produce lean carcasses, because few cooks like to drain a bunch of grease off their meat. Tenderness is also important, so you should look for breeds — or even certain lines within breeds — that are known for their eating quality. (If the phrase "lean and tender beef" sounds like an oxymoron to you, check out Chapter 15, where we talk more about what it takes to raise cattle that produce great-tasting lean beef.)

- **Mature body size:** This characteristic refers to the weight and height of an adult bovine. Larger mature body size of parent cattle normally results in larger calf size, which could lead to birthing difficulties. However, these bigger cattle usually produce calves with heavier weaning weights, which is important if you're selling feeders by the pound. An animal with a larger mature size requires more feed for body maintenance, so you want to be sure you have the right climate for producing plenty of feedstuffs. A 1,500-pound cow isn't the best match for an arid climate with meager grass!

- **Milk production:** This measure refers to the amount of milk that a cow produces for her calf. Increasing your cows' milk production increases the weaning weight of their calves. Heavy milking cows often produce milk at the expense of other bodily functions, however. While producing milk, they may lose weight because all of their calories go toward providing milk. They also may be slower to rebreed. If you have ample, high-quality feedstuffs, you can better care for big, heavy milking cows than if your cattle feed is mainly sparse pasture.
- **Rate and efficiency of gain:** This measure shows how much cattle grow over a period of time and how much feed it takes to produce a pound of gain. Growth is often expressed as average daily gain (ADG) and is calculated by dividing how many pounds an animal gains over a set number of days.

  Cattle breeds that have been selected for a high rate of gain (an ADG of 3 or more pounds from weaning to butchering) often require high-energy feeds like grain to reach their full potential. For details on the nutritional value of various feeds, pay a visit to Chapter 5.
- **Environmental adaptability:** This characteristic refers to the ability of cattle to thrive in challenging conditions due to weather extremes, insects, or sparse feed. A live calf with a less-than-stellar ADG or carcass merit is still worth more than the calf that didn't survive because it couldn't handle the environment.

The bottom line when purchasing new cattle to add to your herd is to try to buy cattle that are raised in a climate similar to your own and that have been raised on feedstuffs that are the same as or similar to what you use. Doing so makes the move less stressful for the cattle. Also, make sure that your new purchases produce a product your customers want to buy so you can sell them more easily and for a better price.

Perusing breed purity: Purebred or crossbred

Cattle that have parents of the same breed are called *purebreds.* Cattle that have parents from two different breeds or parents whose breed is unknown are called *crossbreds.* For example, a purebred Polled Hereford has a purebred Polled Hereford sire and dam. A crossbred could have an Angus sire and a Simmental dam or a parent of unknown ancestry. In the following section, we give you the rundown on the pros and cons for both purebred and crossbred cattle.

Opting for purebreds

Purebred cattle are eligible for registry with a recognized breed association, such as the American Angus Association. These organizations keep records and pedigrees of a particular breed, set the breed's standards, promote the breed, and serve beef producers raising that breed through education and marketing efforts.

If you plan to raise and sell breeding stock, you should choose purebred cattle. Even if you ultimately sell some crossbred cattle, you need the purebred lines to serve as your breeding program foundation. If you want to show your cattle at competitions and fairs, you'll find far more options for showing purebred cattle than you will for crossbred animals. (Chapter 14 provides more information about showing cattle.)

In addition, certain marketing opportunities are available only to particular breeds of cattle or their offspring, such as the registered trademarks for Certified Hereford Beef or Certified Angus Beef. Also, many beef producers with crossbred cows want a purebred bull with a known genetic background. By using a purebred bull, they hope to produce a more uniform calf crop and improve specific traits.

If you raise purebred cattle, make the effort to keep the data and records so you can get registration papers for your animals. *Registration papers* are documents showing the identification, parentage, and expected breeding performance of an animal. Cattle need registration papers to be eligible for some shows and sales. Also, many buyers want registration papers because of the information and potential added resale value they provide.

Going with crossbreds

Crossbred cattle are the result of mating animals from different breeds. Crossbreds exhibit *hybrid vigor,* or *heterosis,* which is the ability to excel in performance areas like growth, fertility, and longevity compared to the performance of their purebred parents. Crossbreeding is done to try to take advantage of the best traits of both purebred parents. Crossbreeding can be especially helpful for improving *low heritability traits,* or characteristics not likely to be passed on from one generation to the next, such as reproductive ability, mothering instinct, and environmental adaptability.

An *F1 crossbred* is the first generation offspring resulting from mating two different purebred cattle of different breeds. When an F1 animal is mated with another F1 cross, their offspring are called the *F2 generation.* F1 crosses show more hybrid vigor and uniformity than F2 animals.

Unless you have a specific plan that requires purebred females, it's hard to beat the benefits of a crossbred cow (from two purebred parents). With her increased fertility, maternal ability, and lifespan, she will often raise more (and bigger) calves than her purebred counterparts.

The benefits of a crossbred bull aren't so clear-cut, however. If you breed crossbred bulls with crossbred cows, you may end up with more variation in size and growth of the calves than what you wanted. Sometimes the data for their potential genetic performance isn't as accurate as that of purebred bulls. So if you want to use breeding selection to improve specific areas in your herd, you're probably better off using a purebred bull.

Not all crossbred cattle are created equal. The term *crossbred* can have different meanings for different people. So when buying crossbred cattle, find out whether their parents are two different breeds of purebred cattle or whether the seller is calling them crossbreds because he doesn't know the genetic background of the parents!

Beefing Up Your Knowledge of Common Beef Breeds

A fun part of raising beef cattle is checking out the features and benefits of the many different breeds and figuring out whether they're right for you. In the following sections, we take a look at several categories of beef cattle: the British breeds, the Continental breeds, the American Brahman, the composite breeds, and some heritage breeds.

The United States and Canada have more than 80 different recognized breeds of cattle, and at least 250 breeds exist worldwide.

The breed descriptions in this chapter are generalizations. Sometimes you may find more variation within a breed than you find between breeds. Buying cattle is a big investment in money and time, so do your homework before making any commitments. Read different breed publications; chat with other cattle producers; and visit farms, sales, and fairs to see what breeds are available. And above all, always keep your goals in mind — they'll guide you in your search for the ideal breed.

Bos indicus and Bos taurus: Check out those ears

Bos indicus and Bos taurus are the two subspecies of beef cattle from which all breeds described in this chapter derive.

Bos indicus cattle have the following distinct physical features: lots of loose, wrinkly skin under their necks and bellies; a large hump made of primarily fatty tissue over the shoulders in both bulls and cows; and large, droopy ears. Their ears are so distinctive that they are sometimes described as "eared-cattle" or cattle with "ear." They're also referred to as *Zebu* cattle due to their Indian Zebu breed ancestors. The common Bos indicus breed in North America is the American Brahman (described in the later section, "The American Brahman: The best known North American Bos indicus"). When the American Brahman is mated with another breed, the resulting offspring aren't 100 percent Bos indicus but do exhibit the Zebu characteristics in a more diluted form.

Bos taurus cattle have tight-fitting skin and smaller ears that come directly out of the side of the head without sagging. Bulls (but not cows) have a muscular bulge in their neck close to the head, but no fatty tissue. Common Bos taurus cattle include the British, Continental, and heritage breeds.

Introducing the British breeds

The British breeds, also known as English breeds, originally hailed from the British Isles. They were first brought to the United States between the late 1700s to late 1800s, and they serve as the foundation for the American beef herd.

Compared to the Continental breeds (covered later in this chapter), the British breeds are generally smaller in mature size and reach mature size at a younger age. They're known for their fertility and *calving ability,* meaning they sire calves that are born easily. Although their carcasses yield less lean meat, the meat they provide is well-marbled.

In the following sections, we take a closer look at some of the more popular British breeds.

Angus

Stockmen introduced Angus cattle to the United States from Scotland in 1873. These solid black cattle were a bit of a novelty at the time because, unlike other breeds, they're naturally hornless.

Over time the Angus has definitely increased in popularity, with more than 250,000 registrations taking place each year. The top five states for Angus registrations are Montana, Texas, Oklahoma, Nebraska, and Missouri.

Angus cattle are recognized for their moderate size and early age at puberty. These characteristics mean they have lower maintenance requirements than larger breeds and can be bred at a younger age. They have good *fleshing ability,* meaning they put on weight easily without as much feed as compared to some other breeds. Because they're smaller at birth, Angus-sired calves are born more easily than larger calves. Additionally, their carcasses have very good marbling. Angus breeders have really capitalized on this marbling trait to promote and "brand" Angus beef as a great eating beef. Figure 3-1 shows an Angus bull.

Hereford/Polled Hereford

Stockmen established the first Hereford breeding herd in the United Sates in 1840. It was 37 years later that breeders noticed the occurrence of polled (or hornless) Herefords. For more than eight decades, the horned and polled breeders maintained separate registries, but they merged together in 1995 under the name American Hereford Association.

Figure 3-1: An Angus bull.

Herefords have a distinctive look. Their reddish-brown bodies contrast sharply with their white faces, bellies, and switches (the hair on the end of their tails). This unique color pattern has helped the Hereford breed maintain more genetic purity than other breeds that accept a variety of colors. This genetic purity means that by using Herefords to produce a crossbred, you get an offspring that has an extra boost of hybrid vigor. (Refer to the earlier section "Going with crossbreds" for details on crossbreds.)

Herefords are known for their quiet dispositions, making them an easy breed to work with and handle. They're hardy animals with good longevity, resulting in many years of productivity. Additionally, they're recognized for having excellent feed conversion, meaning they require less feed to produce a pound of gain, and high fertility, meaning that they are quick to rebreed and give birth on a regular schedule. Figure 3-2 shows a Polled Hereford cow.

Red Angus

The Red Angus shares many traits with its black Angus relative (see the earlier section "Angus") except for, of course, the color. Red Angus cattle are a striking brownish-orange-red color from head to tail.

Figure 3-2: A Polled Hereford cow.

Their high-quality carcasses produce a well-marbled meat, while their comparatively smaller size translates into lower maintenance requirements. Because they're smaller, Red Angus-sired calves are born more easily and require less assistance at calving time. And speaking of calves, Red Angus are attentive mothers who take good care of their young.

Shorthorn

The original Shorthorns came to the United States in 1783 and were valued as much for their milking abilities as they were for their beef-producing abilities. In the mid-1800s, stockmen began importing the breed directly from Scotland, and the breed transitioned toward being more for meat production. However, in the dairy industry, the Milking Shorthorn breed is still in existence today.

Just as with the Hereford breed (discussed earlier in the chapter), the original Shorthorns were all horned, but naturally occurring polled cattle were discovered later. Currently both Horned and Polled Shorthorns are available. More than 18,000 Shorthorns are registered annually, with Iowa, Illinois, Indiana, Ohio, and Texas topping the number of registrations.

The hair coats of Shorthorns can vary significantly. They can be pure white or solid dark red, or they can be any combination of those two colors. Shorthorns don't have large spots of color, though. Instead, the red and white hair closely intermingles in a pattern called *roan.* When Shorthorns are crossbred with black-haired cattle, the resulting offspring may be blue roans. The mixing of the black and white hairs gives the animal a distinctive blue cast.

With a nod to its multipurpose background, the Shorthorn is known as a strong milker, resulting in good growth of the nursing calf. They're a docile breed, which means that they're easy to work with and handle. Shorthorns are attentive mothers and are known for their reproductive abilities.

A well-cultured cow: Looking at the Continental breeds

The Continental breeds are also known as European breeds and sometimes as "exotics." Relatively speaking, they're new on the United States cattle scene, with the first large groups of imports arriving in the late 1960s and early 1970s (except for the Texas Longhorn). The Continental breeds originated from various countries of mainland Europe.

Compared to the English breeds, exotics are known for larger mature size, reaching mature size at an older age, and producing carcasses with less fat but also less marbling. The Continental sires are more prone to cause calving difficulties than their British counterparts.

Just as trends exist in fashion, popular entertainment, and a variety of other pursuits, livestock breeding follows fads, too. In the 1950s and 1960s, cattle producers put an emphasis on making the British breeds in the U.S. shorter and wider in stature. However, with this focus on reducing the mature animal size, the British breeds' growth ability and lean carcass yield were decreased as well. Cattle producers soon began looking for breeding options that would put some growth and muscle back into their cattle, and the Continentals fit the bill.

Due to the great demand for, and limited number of, Continentals originally available in the United States, most Continental breed associations allowed *breeding up.* This practice starts by mating a purebred of the European breed with another breed. The resulting offspring are then mated to another purebred of the same European breed. After four or five generations (depending on the rules for each breed), the animal produced could be called a purebred. Breeding up helped to quickly increase the population of exotics and introduce new traits for the breeds, such as black color and being polled.

In the following sections, we review several of the more notable Continental breeds.

Charolais

The Charolais is unique among its exotic companions for two reasons. One reason is its distinguishing solid white to straw-yellow color with light-colored muzzle and pigment. Also, while all the other exotics (except the Longhorn) entered the United States for the first time in the 1960s or 1970s, the Charolais first arrived here from France, via Mexico, in the 1930s.

The Charolais is a large animal with heavy muscle, meaning it produces large, lean carcasses with a good yield grade. This breed is also an efficient converter of high-energy feed into pounds of meat.

Chianina

Very few full-blooded Chianina (kee-ah-neen-a) cattle live in the United States or Canada. When you see them though, they're recognizable by their wiry white to steel gray hair and black noses and pigment. The breed was founded in central Italy and is considered the largest beef breed, with males reaching 3,000 pounds.

With their exceptional growth rate and high lean to fat carcass ratio, they make complimentary crosses with the British breeds of cattle. In fact, the American Chianina Association not only registers full-blooded and purebred Chianinas but also Chiangus (Chianina X Angus), Chiford (Chianina X Hereford), and Red ChiAngus (Chianina X Red Angus) animals. Chianina crosses are often quite competitive in steer shows at the state and national level.

Unlike most of the other Continental breeds, Chianinas are known for their ease of calving.

Gelbvieh

This breed came to the United States from Germany in 1971. The Gelbvieh (gelb-vee) can be any color and can be horned or polled. They rank seventh in the number of registrations among beef cattle in the U.S.

The Gelbvieh breed strives for a balance of traits that are usually just associated with either the British or Continental breeds. They're recognized for fertility and milking ability, while also producing calves that are well-muscled and grow quickly.

Limousin

Limousin (pronounced just like the car) cattle are muscular animals that originated in France and came to Canada in 1968 and the United States in 1971. They range in color from golden red to black. This breed can produce lean carcasses in a variety of environments while growing rapidly and using feed efficiently.

Limousin cattle have had a bad reputation regarding their temperament, but breeders are working to assess and improve the dispositions of their animals.

Maine-Anjou

Like the Limousin, the Maine-Anjou breed originated in France. The breed's genetics first came to the United States in the form of frozen semen imported from Canada in 1969. The foundational Maine-Anjous were dark red with white on their heads, bellies, rear legs, and/or tails. However, through the process of breeding up, many of today's Maine-Anjou are black and polled.

This breed has a large mature size with bulls weighing from 2,220 to 3,000 pounds and cows weighing more than 1,500 pounds. Maine-Anjous are known for their ability to produce lean, muscular carcasses.

Saler

The Saler (sa-lair) breed originated in France and first arrived in the United States in 1972. The first domestic Salers were horned and had deep, dark red coats, but with breeding up, more are becoming polled and black.

Salers deviate from the norm of the Continentals in several ways: They're known for calving ease and for their mothering ability. Additionally, they show more marbling and can handle rough terrain and harsh weather. This breed has had problems with wild dispositions, but over the last few years breeders have been focusing on producing more docile animals.

Simmental

Although this breed originated in the Simme Valley of Switzerland, it has become popular worldwide with 40 to 60 million head across the globe. Around 50,000 Simmentals are registered annually in the United States. The first Simmental arrived in North America in 1967. Over time the color of the breed has evolved from light yellow or red with white spots to gray and now primarily black or dark red with minimal white markings.

Out of all the Continentals, the Simmental is recognized as the heaviest milker. This high milk production results in large calves at weaning time. The growth continues as the animal matures, allowing the Simmental to produce large carcasses. SimAngus (Simmental X Angus) cows are a popular choice for crossbred cow herds (groups of cows). Figure 3-3 shows a Simmental bull.

Figure 3-3: A Simmental bull.

Texas Longhorn

These unique animals are descendants of Spanish cattle brought to North America by Christopher Columbus. They have large, long horns and come in a variety of colors and color patterns. They're widely recognized as an iconic figure of the Wild West. Because they lived as feral cattle (domestic animals that have turned wild) for more than 300 years, natural selection has made them hardy and long-lived. They have a tendency to be light muscled and to produce small calves.

The American Brahman: The best known North American Bos indicus

The first Bos indicus cattle arrived in America in 1854 from India (for a little background on Bos indicus, see the sidebar "Bos indicus and Bos taurus: Check out those ears" earlier in this chapter). In subsequent years, Bos indicus were also imported from Brazil. Even though more than 30 strains of these "Indian cattle" came to the U.S., only a few were used to create the beef cattle that would go on to become the American Brahman.

Brahman cattle are hardy animals. They can live on low-quality, sparse feed and are able to resist insects and external parasites (more on those in Chapter 10). Brahman continue to grow and breed even in hot weather that would negatively impact the performance of Bos taurus cattle. However, buyers of feeder, stocker, and fat cattle significantly reduce the price paid for Brahmans and cattle showing Brahman characteristics like the droopy ears and loose skin. This is because Brahman meat is so lean that it can be less juicy and tender than other breeds. In the southeast, south, and southwest United States, Brahmans are often mated with other breeds to produce "Brahman-influenced" offspring that have the Brahman's ability to resist heat and bugs while retaining the eating quality derived from Bos taurus bloodlines.

If your climate has unrelenting heat and humidity, plus lots and lots of bugs with limited feed resources, cattle with some Brahman genetics could have a place on your farm. Just realize Brahman-influenced cattle may be worth less pound for pound than beef cattle without "ear."

Maintaining vigor with the composite breeds

Composites refer to cattle made of two or more breeds. These cattle are designed to maintain hybrid vigor in future generations without crossbreeding. All four of the main composite breeds — Brangus, Beefmaster,

Santa Gertrudis, and Simbrah — in the United States were developed using Brahman cattle crossed with British or Continental breeds. Each of the composites is recognized for its hardiness and ability to withstand heat. Table 3-1 shows the different breeds, their percentage of foundational breeds, color, and horned/polled status.

Table 3-1 Composite Breeds

Breed Name	*Foundation Breeds/ Percentage*	*Color*	*Horned/Polled Status*
Brangus	⅝ Angus, ⅜ Brahman	Black	Polled
Beefmaster	½ Brahman, ¼ Hereford, ¼ Shorthorn	Red or light brown	Mainly horned
Santa Gertrudis	⅝ Shorthorn, ⅜ Brahman	Dark red	Polled or horned
Simbrah	⅝ Simmental, ⅜ Brahman	Variety of colors	Polled or horned

Heritage breeds: Examining the old, pure bloodlines

A *heritage breed* is a breed that has been in the United States for many years and that is *genetically true,* meaning that the breed traits are consistently passed from one generation to the next. Heritage breeds also have limited registrations and a small global population. Here's a brief overview of some of the most common heritage breeds found in the U.S.:

- **Belted Galloway:** Sometimes called Oreo Cookie cattle because of their black fronts and ends and white middles, Belted Galloways have small to moderate frames and are well-adapted to cold climates. The Belted Galloway is an efficient converter of forage into meat.
- **Devon:** These dark red cattle were some of the first cattle brought to the United States. They're hardy animals with good grazing ability.
- **Highland:** These unique-looking cattle with shaggy orange/brown hair and long horns do well in cold, wet climates and can utilize rough forage.
- **Red Poll:** The cattle in this fertile breed are moderate in size. As their name indicates, they're dark red in color and don't have horns. Red Polls are known for their ability to use grass for weight gain.

How we chose the cattle that roam the pastures at Royer Farm

When we were choosing the cattle for our farm, we started with a goal: to raise beef and sell it directly to the consumer. So our ultimate target was to produce animals with lots of lean beef (high yield) while having great eating quality. We have a grass-based farm with some of our market cattle supplemented with corn, whereas others are 100-percent forage fed. Our plan to raise lean beef primarily on pasture led to the need for moderately sized animals that would have low body maintenance requirements and good fleshing ability.

To achieve this combination of traits, we mate a purebred bull (either Angus or homozygous polled Simmental) with our predominately F1 crossbred cow herd. The cows are a combination of Angus X Charolais, Angus X Simmental, and Red Angus (these are some of the most common breeds in our area). The reproductive and maternal hybrid vigor of a crossbred cow means that more cows have a calf and that the calves grow better. The Angus bloodlines provide moderately sized animals and a good eating quality while the Simmental and Charolais efficiently produce lean beef.

We purchase any herd additions from farms within 150 miles of our place. As a result, the cattle are accustomed to our cold Indiana winters and hot, dry summers. Also, we only purchase cattle that were comfortable grazing for their feed and didn't expect a person to feed them.

Part II

Preparing to Bring Home the Beef

"You gonna get that cow shelter built soon?"

In this part . . .

This group of chapters helps you get off to a good start with your cattle. Chapter 4 walks you through all the facility and fence options you need to consider. In Chapter 5, we provide the skinny on the nutritional aspects of caring for cattle. Chapter 6 shows you how to properly manage your pastures, manure, and water in harmony with your cattle. Finally, when you're done preparing, it's time to go shopping! Chapter 7 steers you to the best places to purchase cattle and tells you what to look for when selecting animals.

Chapter 4

Home on the Range: Preparing Your Facilities

In This Chapter

- Installing secure fences and gates
- Creating effective handling facilities
- Providing shelter, protection from the weather, and bedding
- Keeping the barnyard clean with HUAP zones

Bovine housing doesn't need to be fancy, but a well-planned farm and fence design helps keep your cattle more content and makes caring for them easier and safer. The required living accommodations are straightforward; your cattle need fenced spaces to graze, a strong pen or enclosure, and clean places to rest, eat, and get away from extreme weather.

This chapter explains the most important aspects of external and internal fencing, safe and secure corrals, and well-designed shelters. We also give you pointers on how to maintain a tidy home for your cattle by creating a HUAP, or heavy use area protection, zone.

Making (and Keeping) Good Neighbors by Building Strong Fences

Strong, well-built fences and strategically placed gates keep your cattle safe and give you peace of mind. Getting a late-night phone call from a neighbor down the road notifying you that your cows are in their garden is a surefire way to ruin your night and possibly neighborhood harmony.

You need two types of fences for your cattle:

- **External or permanent fences:** These fences keep your cattle from roaming off your property or into valuable crop fields. They're generally built to be in place for at least seven years or more.
- **Internal or subdivision fences:** These fences help you keep your cattle restricted to certain portions of your land as needed. They're important because the better control you have over your cattle's grazing the more you can manage their consumption and use of pasture. These fences are typically designed to be in place anywhere from one day to several years.

We provide a detailed explanation of these two types of fences in the following sections. We also help you determine what kinds of gates are best for your fencing and provide some advice on electrifying your fencing. For more information about the amount of pasture your cattle need, see Chapter 6.

Addressing permanent fence needs

Permanent or external fencing is great for providing your cattle with clear boundaries so they stay put — and stay safe.

If you have an existing fence on your property, check it carefully to ensure it's ready for holding cattle. Refer to the guidelines in this section to confirm that the construction and materials are of a high quality.

If your pastures already have fencing, but it's weak or broken down, you can patch it. For isolated problems, use a panel or gate to block a hole. If the entire fence needs replacing, run a strand of electrified wire 12 to 18 inches inside the external fence until you have a chance to build a new permanent fence.

If you don't have existing fencing, or your existing fencing is in need of repair, you need to decide whether to build or repair it yourself or to hire someone else to do it. We've put up our own permanent fencing, and we've also hired builders to install it for us. Both situations have pros and cons.

You need the knowledge, time, equipment, and physical strength to do the job properly yourself. If you decide to go with a custom builder, get references and actually check them. Examine jobs that are five or ten years old to make sure the builder's fences stand the test of time.

Woven wire fencing

A high-tensile woven wire fence with a single high-tensile wire strung above the woven wire and an offset electrical wire on the pasture (not the public) side of the woven wire is hard to beat as a permanent fence, even though it's one of the more expensive fence designs. (*Tensile strength* refers to how much force is required to pull something like a wire before it breaks. Most fencing is made of steel wires, but the high-tensile variety has more strength than the regular, traditional steel fencing.) Refer to Figure 4-1 to see what this type of fencing system looks like.

This configuration can contain all classes of cattle, and it gives you the flexibility to allow other species, such as horses, hogs, or sheep, to use the same pasture as the cattle.

The most important parts of a woven wire fence system are as follows:

- **Woven wire:** High-tensile woven wire serves as the main physical and visual barrier. It should be at least 48 inches tall, with the bottom of the fence about 3 inches off the ground. When using galvanized high-tensile woven wire, 12.5 gauge (g.) is sufficient. Wires are sized by gauge, with smaller numbers being larger in diameter and hence stronger.

  Fence manufacturers describe woven fence using a shorthand system that looks like this: ##/##/##. The first number indicates the number of horizontal wires, the second number indicates the top-to-bottom height of the fence, and the third number indicates the spacing of the vertical wires. So a fence labeled 9/49/12 has 9 horizontal wires with a top-to-bottom height of 49 inches, and the vertical wires are spaced every 12 inches.

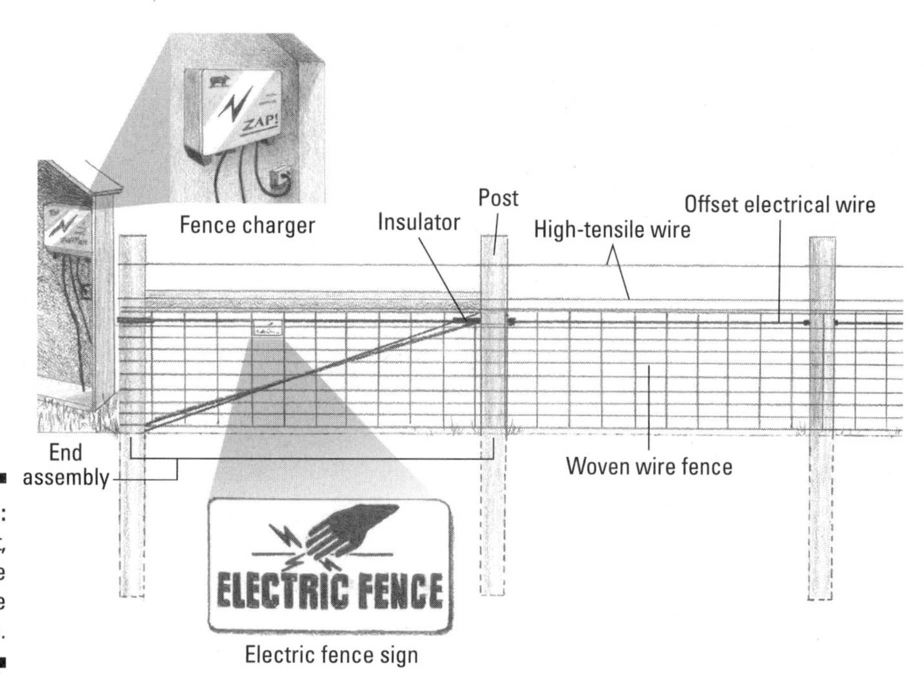

Figure 4-1: A permanent, woven wire fence system.

- **High-tensile wire:** This wire serves as an additional barrier to taller animals. This single wire is springy and can be retightened or replaced more easily than woven wire. So if a tree limb or other object falls on it, the wire will bounce back after the weight is removed, and your expensive and less flexible woven wire is protected. String the single 12.5 g. high-tensile wire strand 4 inches above the top of your woven wire.
- **Offset electrical wire:** To keep animals from pushing on your woven wire fence, you can string another individual strand of 12.5 g. high-tensile on the inside of the fence posts 2 feet above the ground. This wire should be hooked up to an electric fence charge so the "hot" wire will carry an electrical current (and you can use the wire to electrify your internal fences, too). See the later section "Keeping cattle in line with electric fencing" to find out more about using electricity to your benefit.
- **Posts:** The best wire fence is useless if it isn't attached to sturdy, long-lasting posts. In most parts of the country, wooden posts are more plentiful and less expensive than steel. However, keep in mind that if you use wooden posts, they need to be pressure-treated with preservatives or made from rot-resistant wood (like Osage orange).

  Your posts in the fence row should be 2 to 2½ feet deep into the ground and be at least 4 inches in diameter at the thick end (but preferably 5 inches for longer fence life). If your fence is on fairly level ground, you can place the posts one rod apart. (A *rod* is a unit of measurement used in construction that's equal to 16½ feet in length.) Hilly terrain usually calls for more frequent post placement.
- **End assemblies:** End assemblies, which are critical to the strength and longevity of your fence, serve as the primary anchor for your entire fence. The assemblies consist of corner posts, brace posts, and high-tensile brace wire. (See Figure 4-2 to get a look at a fence with a proper end assembly.)

  The corner post should be set 3½ feet in the ground and should be at least 6 inches in diameter at the narrow end. Both the in-ground brace post and horizontal brace post need to be 5 inches in diameter. The brace wire made of high-tensile wire needs to double loop around the brace posts in a figure-eight pattern. For lines of fences greater than 200 feet, you should use a double-span end assembly for extra strength. A *double-span end assembly* is a corner post supported by two consecutive sets of brace and horizontal posts instead of just a single brace and horizontal post.

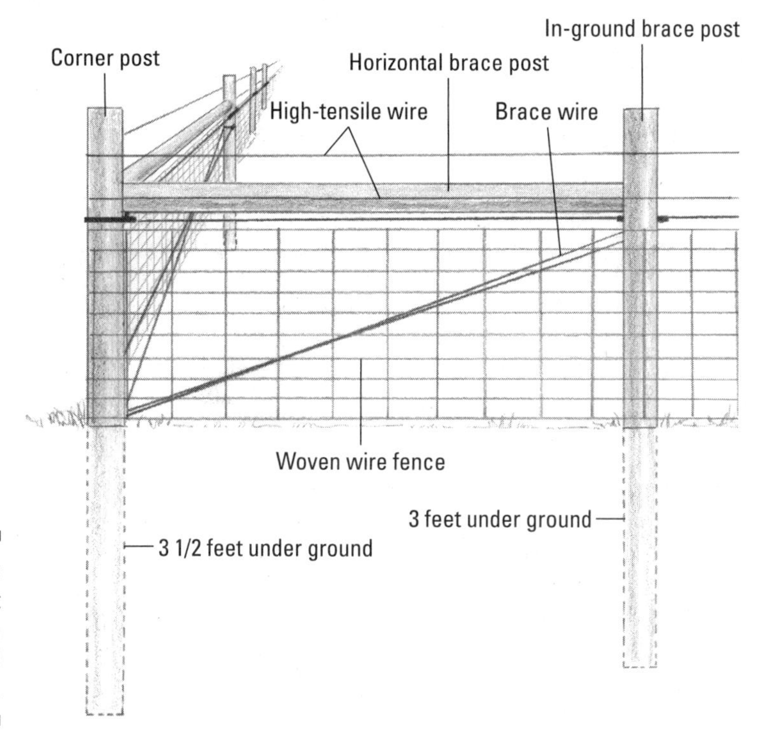

Figure 4-2: A permanent fence with a proper end assembly.

Barbed wire fencing

Barbed wire fencing is one of the cheaper fencing options, which makes it especially helpful when you need to cover large distances in remote rural areas. However, if you plan to keep your cattle by a frequently traveled road or close to crop fields and homes, barbed wire isn't a good fence choice because even six strands of tightly strung barbed wire may not be enough to contain a hungry cow or an adventurous calf.

You should only use barbed wire around pastures where only cattle live because it's dangerous for horses and ineffective for sheep.

If you have barbed wire fencing, keep it in good repair and firmly attached to posts. Loose barbed wire can become tangled around a cow's legs or neck, causing severe injury or death. Also, for safety reasons, never electrify barbed wire fences because an animal or person can become caught by the barbs and unable to escape the electrical pulses (see the later section "Staying safe with electric fencing" for more on the proper use of an electrified fence).

Preparing internal fencing

If you've invested lots of time and money into your permanent fencing system, you may not want to deal with another post or roll of wire ever again. But we encourage you not to give up on your fencing projects! Internal fencing allows you to more closely manage the grazing habits of your cattle because they will more uniformly eat a smaller section of grass than if they were allowed to meander all over your pastures. The benefits of using internal fencing to subdivide your pastures include increased grass production and cleaner living spaces for your cattle.

Luckily, internal fences are easier to build than permanent fences, and they're less expensive, too. You can build these fences yourself or hire someone to build them as your time and budget allow.

You can use these two basic types of fences inside your fields:

- **Semipermanent fences:** These fences are best for dividing your larger pastures into smaller paddocks. They can last for several years and can be made of different types of materials. For cattle, we've found that two strands of electrified wire held in place by fiberglass posts makes a sufficient semipermanent fence. This type of system prevents the need for a full-fledged end assembly. These fences are easy to set up; you can use a single large wooden post at the corners of your paddock and 4-foot-long fiberglass posts tapped into the ground along the edge of the pasture.

  Don't depend on this fence design to keep cattle out of yards, off roads, or away from grain fields because they're not substantial enough to stop a very determined animal.

- **Polywire conductor fences:** These fences are the best choice when your boundary needs to be moved daily or weekly. The two parts of this fence are polywire and step-in posts. *Polywire,* also known as *electric twine,* is made of interwoven strands of metal and plastic. The metal conducts electricity, and the plastic provides strength and flexibility. You use lightweight plastic posts with clips to hold the single polywire in place. You can push these posts into the ground by pressing your foot onto the step-in portion of the post. With a little practice, you can install and move this fence about as quickly as you can walk.

Planning for gates

Gates provide an opening so cattle can move safely from one fenced area to another. Gates can range from a simple wire panel attached to the fence post with baling wire (we have plenty of those on our farm) to heavy-duty steel or wooden board gates that swing smoothly on hinges. Make sure your gates are at least as strong and durable as your fence so they aren't the weak link when it comes to containing your cattle.

To build or not to build

Although building external fencing can be daunting, tackling internal fencing on your own is much easier. You don't need as much equipment or brute muscle, and you gain experience for bigger projects in the future. If you have a limited fencing budget, putting up your own internal fence saves you money by cutting down on labor expenses. You'll also experience a satisfying feeling of accomplishment when you look over a long stretch of fence that you installed. Fencing websites and catalogs from companies such as Kencove (`www.kencove.com`) and Premier (`www.premier1supplies.com`) are great resources full of products and advice.

If you add new fences to your property, put gates at the corners of your pasture. Cattle tend to stop moving and cluster in corners of a field, so gates in those locations are easier for them to find. Also, if you put gates at the end of a fence run, you don't need to build additional end assemblies, which means you save time and money.

Here are some tips for getting the most out of your gates:

- Make the gates you use often the easiest to operate; doing so makes daily chores quicker and more enjoyable.
- Consider what you want to pass through the gates (besides cattle) and size them accordingly. For example, 12-foot-wide gates allow trucks to pass through easily, but they aren't wide enough for large farm equipment.
- Make sure the bottom of your gate is 6 inches off the ground to account for some sagging and soil buildup over time. To help prevent some of the sagging, prop the end of the gate opposite the hinges on a wooden block or metal peg driven into the adjacent post.
- To prevent the gate from being lifted off its hinges by cattle rubbing on it or by two-legged troublemakers, place your bottom hinge upward and your top hinge pointing downward (see Figure 4-3).

After you determine what type of gate you'll use, don't forget you have to make decisions on the latch as well. The latch design on your external and internal gates should be secure and easy to use. A heavy-duty chain and padlock keeps unwanted vehicles out of your pasture and is what we use.

For our internal fencing, we secure our gates with light welded chain that's attached to the gate. Because the chain is part of the gate, it can't "disappear" like a strand of wire or rope. Tuck the end of the chain under its loop so a cow can't accidentally open the gate if she plays with the dangling end!

Figure 4-3: Helpful placement of hinges on a gate.

Keeping cattle in line with electric fencing

Electric or energized fencing is made of one or more strands of wire that have an electric charge going through them. When cattle touch the fencing, they get a sharp zap that makes them want to stay away from it.

The key component of your electric fence system is the *energizer* or *charger.* Energizers are about the size of a big shoe box and take power from a regular electrical outlet, a battery, or the sun (via a solar panel); they convert the power into electrical pulses that travel through the electric fence.

You can choose from four types of fence chargers: 110-volt AC plug-in energizers, DC battery energizers, AC/DC battery/plug-in energizers, and solar energizers. Here's the lowdown on each:

- **110-volt AC plug-in:** These chargers can be powered to carry a heavier load and electrify more miles of fence than other chargers. After the initial purchase price (which can range from $100 to more than $600 dollars for a powerful charger) they're inexpensive to operate and maintain because they can be plugged into a standard electrical outlet and don't have pricey batteries like DC and solar units. We find these chargers to be less susceptible to lightning damage, too. We recommend these over the other chargers.

- **DC battery energizers:** To create a powerful pulse you need to use a 12-volt battery, and even then these units can't charge as much fence as a 110-volt AC plug-in charger. You also need to remember to recharge the battery.
- **AC/DC battery/plug-in energizers:** These can be powered by either plugging into an electrical outlet or with a 12-volt DC battery. These units give you good flexibility because you can choose between two power sources. However, due to the battery being outside the unit, they can be cumbersome to install when using the AC power option.
- **Solar:** These chargers are simple to install. However, even high-dollar solar chargers tend to be weaker than 110-volt AC units and more prone to lightning damage.

Cattle won't want to touch an energized fence so they're less likely to rub or crowd the fence, break it down, or escape. After cattle become accustomed to electric fencing, you can use inexpensive single strands to contain them in your internal paddock subdivisions.

In the following sections, we help you stay safe with electric fencing, and we show you how to keep up the performance of the product you choose.

Staying safe with electric fencing

You're smart, so we don't have to tell you that electric fencing can be dangerous to you, your animals, and unsuspecting neighbors and visitors. But we do want to give you the basics on keeping your farm safe for humans and animals.

You should place warning signs along your stretches of fence so people know that it's electrified. When possible, put the conductive wires inside the perimeter fence so passersby can't touch the wire accidentally. If human contact is possible, look for a charger with a short and low initial pulse energy (60 seconds). This initial mild pulse gives the person or animal time to move away before really getting zapped.

Here are some guidelines to follow when using electric fencing:

- **Never connect an electrical fence directly to an electrical outlet.** The constant current flowing through a household or barn outlet is dangerous compared to the quick electrical pulses released by a fence charger.
- **Take care to never contact the wire with your head or neck.** Your delicate nervous system is more exposed in these areas of the body.
- **Never let unsupervised guests or children around any electrified fences.** Even if you warn them about the discomfort and potential danger caused by electrical fence, they may forget and accidentally touch the fence.

- **For safety reasons, never electrify barbed wire.** An animal or person can become caught by the barbs and unable to escape the electrical pulses.

Setting up your fencing and keeping its performance strong

Of course, if you're going to invest in electric fencing, you want to make sure you use it to its fullest potential. As you're setting up your fence, be sure to follow these tips for top performance:

- **Purchase an adequate charger.** If you're like us, after you have your electric fence system up and running, you'll be so happy with it that you'll want to use it for more and more fence. If the fence isn't exposed to the public, a good starting unit should have a 5- to 6-joule output.
- **Train animals to respect the electric fence.** When animals are accustomed to a woven wire fence, a little bitty hot wire may not register as a barrier to them. To get animals used to an electric fence, set up an electrified training fence inside a permanent fence, and then lure the animals up to the energized wire by placing hay or grain on the other side of the fence. This training enables the cattle to come in contact with the electric fence in a controlled environment.
- **Use your voltmeter.** Don't go through the work of setting up an electric fence without checking to see whether it's actually electrified! Test your fence using a voltmeter specifically made for checking electric fences. A charge of 5,000 to 6,000 volts results in a safe but robust zap to cattle.

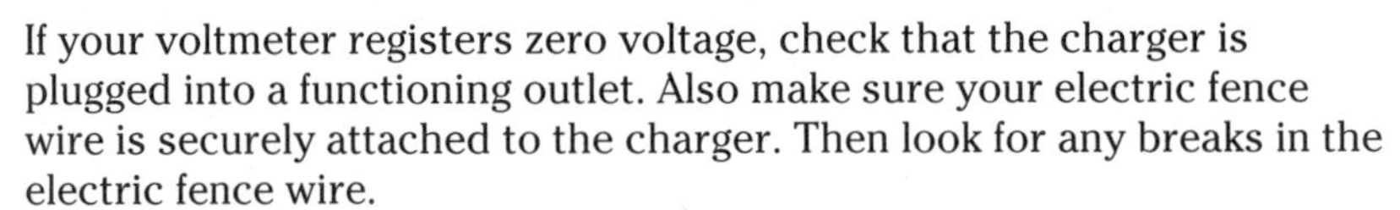

If your voltmeter registers zero voltage, check that the charger is plugged into a functioning outlet. Also make sure your electric fence wire is securely attached to the charger. Then look for any breaks in the electric fence wire.

Speaking of voltage, the manual that comes with your charger should detail the expected voltage power and the distance of fence it can energize. If your charger isn't performing up to its potential, here are three possible problems:

- **You have grass or brush growing on your fence.** Green, lush grass and brush drains the "juice" out of an electric fence. Trim or use herbicide on your fence row to keep the vegetation at bay.
- **You don't have sufficient ground rods.** Your ground rod can be a copper pipe or galvanized steel rod (galvanized to prevent rust) that you drive or hammer into the ground. Don't skimp on ground rods because your fence and charger won't operate to their full capacity without proper grounding. Make sure you have at least 3 feet of ground rod in the soil for every joule of energy released from your energizer. For example, a 5-joule energizer requires 15 feet of ground rod. Be sure the ground rods are securely connected to the charger's insulated ground wire.

- **You have an electrical short.** An electrical short occurs when the electricity can't flow smoothly through the electric fence. When the electricity meets resistance, the voltage drops. Causes of resistance include rubbing against a post or vegetation, or partially intact or frayed "hot" wire.

Often at the site of the short, the fence will give off a snapping noise with every electrical pulse. If you can't hear where the problem is, you have to visually inspect the fence. After you've found the short, move the fence away from the physical barrier or cut out the frayed wire and splice the fence back together.

Planning Your Handling Facilities

Your *handling facility* consists of groups of pens, alleyways, and equipment that provide a place for you to care for your livestock safely, properly, and in a low-stress manner. Basic handling facilities include a sturdy corral to catch your cattle in and a head gate and chute to control your cattle.

Cattle handling facilities may seem like a big expense for something you may only use a few times a year, but you'll be happy to have them when you need them. Inadequate working areas make it difficult to do basic animal preventive health maintenance and to care for your cattle in case of an emergency.

To get started with your handling facilities, you need to select and design your site. Then you're ready to choose your equipment. The following sections outline everything you need to know.

Selecting a site for your handling facilities

To make handling your cattle as simple and stress-free as possible, you want to locate your handling facilities in a good spot. When selecting a site for your facilities, choose a location that is

- **Accessible to cattle, people, trucks, and trailers in all kinds of weather, including rainy and snowy seasons:** You may need to work with or transport your cattle even if the ground is a sticky, muddy mess or covered with 2 feet of snow.
- **Convenient to numerous pastures:** A good spot is the corner of a pasture where several paddocks converge, especially if the pasture has strong fences.
- **Well-drained:** A muddy mess is difficult to work in and is stressful to the cattle.

- **Close to utilities, such as water and electricity:** For example, you want to provide your cattle with water, especially on hot days or if they'll be penned overnight. And don't forget that emergencies don't just happen during daylight. So having good, permanent lighting makes for better operating conditions than working under the glare of truck headlights!
- **Considerate of your neighbors:** You may use your handling facilities at any time of day or night and your neighbors may not appreciate all the commotion. Plus, dust and flies aren't welcome houseguests.

Designing your handling facilities

After you've selected your handling facility site (see the preceding section), you're ready to start designing the layout. Be sure to include the following main parts in your design (they're listed in the order through which the animals move through your facility):

- **Holding pen:** This pen allows you to gather a large group of cattle in one spot.
- **Crowding pen:** This pen allows animals to be confined in a smaller space and to be made to move in the needed direction.
- **Working chute:** This narrow, solid-sided alleyway helps you line up the cattle single file in a nose-to-tail fashion.
- **Loading chute:** You use this ramp or alley to get cattle from the crowding pen onto a trailer.
- **Head gate:** This piece of equipment has two metal panels that squeeze down on the cow's neck to prevent the animal from moving forward or backward. Head gates can be either automatic (self-catching), where the pressure of the animal's shoulder on the panels causes the panels to squeeze in on the neck, or manual, where a person pulls down on a lever to close the panels.
- **Holding chute:** This chute is a big metal cage attached to the head gate. It prevents the animal from moving from side to side while in the head gate.

Figure 4-4 shows a basic layout that includes the preceding components.

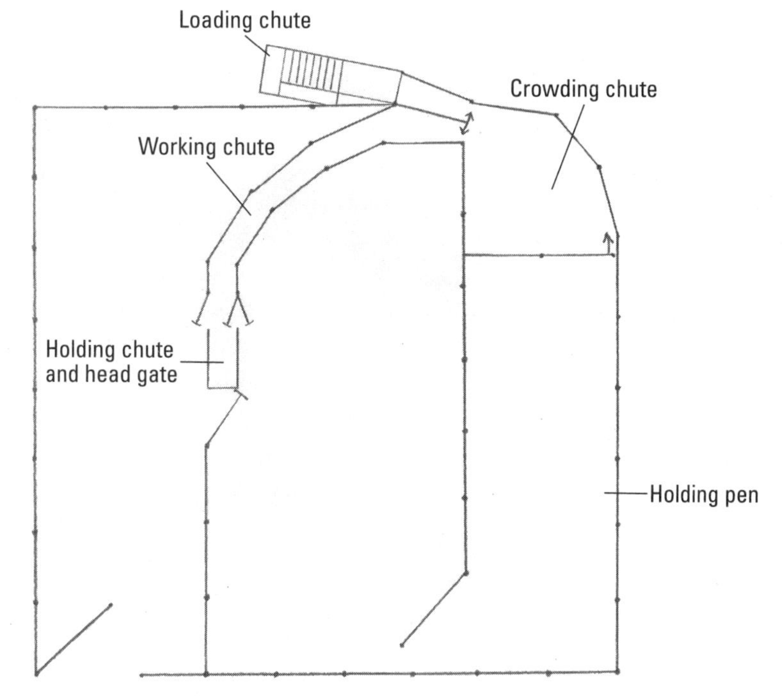

Figure 4-4: A basic cattle handling facility layout.

Here are some factors to keep in mind when designing the different parts of your handling facility:

- **Strength:** Even if your cattle are tame and quiet, sometimes they will still test the strength of your facilities (such as when you separate a protective mother from her calf). To make sure you have a strong facility, follow these guidelines:
    - Make sure you place all the posts used in the pens and chutes 2½ to 3 feet in the ground and no more than 6 feet apart.
    - The sides of your pens should be made from sturdy materials like steel gates, cattle fence panels reinforced with three 2-x-6 wooden boards (at the top, middle, and bottom), or solid sheets of metal welded onto a strong pipe frame. The sides should be a minimum of 5 feet tall.

- **Space:** Providing the proper space in your facility is important for safety and ease of use. Too much space allows the cattle to run around nervously, and too little space makes it difficult for the cattle to move where you want them to go. Here we provide our space recommendations for each of the facility areas:
  - For your holding pen, plan on 20 square feet per mature animal or 35 square feet for a cow and her calf.
  - The alleyway leading from the holding pen to the crowding pen or loading chute should be 10 to 12 feet wide. Cattle are less likely to bolt around you with this size alleyway, but they still have some space to get around you instead of running you over if they really want to flee.
  - The crowding pen is a small space made to hold eight to ten head with 10 square feet per cow. Commercially available circular crowding tubs work nicely to get cattle moving quietly and safely either to the loading chute for transport or into the working chute.
  - Your working chute should be 22 to 26 inches wide at the top and 20 feet long.
- **Ease of use:** Take advantage of your cattle's natural tendencies to make your facility easier to use. For example, consider doing the following:
  - Put gates in the corners of the holding pens, because cattle like to congregate there.
  - Use solid-sides on your crowding pen, working chute, and loading chute so the one exit the cattle see takes them where you need them to go.
  - Curve your working chute to capitalize on the desire of cattle to skirt around their handler and to circle back to where they came from.
- **Safety:** When working with cattle at any time, but especially in the close quarters of a handling facility, make safety a top priority. Follow this advice:
  - If you use concrete flooring, score it with deep parallel grooves or a diamond-shaped pattern to prevent slips and injuries to humans and bovines alike.
  - Install gates that latch shut automatically to prevent an animal from knocking the gate open and hitting a person or another animal (this advice is especially important in the tight spaces of the crowding pen).
  - Install safety passes or safety gates in several places in your facility to provide a quick escape route for people. A *safety pass* is a 14-inch gap in the fence with sturdy fence posts on each side. A *safety gate* covers this opening with a spring-loaded gate for situations where a small calf may try to squeeze through the opening.

Don't use barbed wire or electrical fence in your handling facility. These barriers rely on pain as a deterrent, and a nervous or angry animal is too likely to ignore the pain and charge through them, endangering the animal and unsuspecting humans in the area.

Selecting a head gate and holding chute

The head gate and holding chute (see Figure 4-5) are important because they allow you to restrain your animal safely so you can perform routine health maintenance or provide emergency care.

When purchasing a head gate and holding chute, consider the following:

- The head gate and chute should operate quietly and smoothly so the cattle aren't frightened by clanging and rattling metal.
- The head gate and chute should be adjustable so the unit can be expanded for your 1,800-pound bull and then squeezed in to catch a young calf.

Figure 4-5: A commercial head gate and chute.

- The holding chute shouldn't be a single solid gate. Ideally it should consist of several smaller gates or openings so you can easily reach the neck for giving shots or safely reach the udder or hooves while keeping the animal restrained.
- You can attach wheels to some gate and chute units so they're portable. If you have cattle in multiple locations, a unit with wheels can be a worthwhile option.

Great places to see and compare many different gate and chute options are at state fairs, farm shows, and livestock exhibitions. You can also look for used equipment at local farm sales or in agricultural classified ads. Just be sure that the head gate and chute are in good and safe working order or that you have the skills to make necessary repairs. A head gate and chute is a big financial investment, so no matter where you buy, find out whether you can take a "test squeeze" before buying.

Most large-animal veterinarians have a portable head gate and chute they can bring to your farm for a nominal fee. If you only need a head gate a few times a year when you're working with your vet, asking the vet to bring hers can be a cost-effective option.

Sheltering Your Cattle

Providing shelter for your cattle is important because it decreases their stress-level, improves their health, and decreases the amount of feed they need. As you start dreaming about a new barn or contemplating remodeling that old tool shed, remember that all cattle shelters need to be

- **Adequately sized:** For shelters that allow the cattle to have access to pasture, a calf needs a minimum 15 to 20 square feet of usable space and each adult needs 20 to 30 square feet. If you will confine animals for any amount of time, you need to provide them with up to 50 percent more barn space. If a big feeder or watering system is in the barn, remember that equipment reduces the usable space. Keep in mind you also need storage space for feed and bedding.
- **Properly ventilated:** The shelter needs windows or doors that you can open to promote air circulation. Adequate ventilation reduces odors and helps the cattle's bedding to stay dry.
- **Well-drained:** Your shelter should be at the top of a hill or on a slope to promote moisture flow away from the barn. By promoting drainage, your cattle will stay cleaner and the wooden parts of your barn will stay drier and last longer.

- **Easy to clean:** If a shelter is difficult to clean, you'll be less likely to tackle this dirty but necessary task. Unless you're a glutton for punishment, the manure and dirty bedding produced by just a few cows will be more than you want to scoop by hand. So consider these questions to determine whether the space will prove easy to clean: Are the doors big enough to drive a skid steer or tractor with loader through? Does the space have support beams you can't drive around? Is the roof high enough that you can lift the scooping bucket? If the space isn't big enough for this helpful machinery, get out your shovel!
- **Safe:** Safety is important if you want to keep your cattle in good shape. So be sure to look around the shelter for any problems. For instance, determine whether the shelter is solid and strong enough to hold up to extreme weather and raucous cows. Also, cattle can get some serious itches and like to rub hard against surfaces. So check for protruding nails and screws that could cause scrapes or wounds. Finally, make sure the floor is made of a nonslip surface and that all electrical wiring is out of reach of curious noses and tongues.

In the following sections, you can read about the different types of shelters that you may use for housing your cattle. We also provide information about shade sources, wind breaks, and bedding.

Going with the tried and true: Pole barns

For the last several decades, new cattle buildings have usually been *pole barns.* These structures are made with steel sides and a roof attached to a wood or steel frame. Pole barns are advantageous because they're

- **Widely accepted:** Because of their wide use and acceptance, materials for pole barns and knowledgeable pole barn builders are fairly easy to find.
- **Customizable:** Pole barns can be designed and configured in many layouts to fit your particular needs.
- **Long-lasting:** Pole barns usually require little maintenance and are long-lasting. One of our main pole barns is into its fourth decade of use with just one new coat of paint and the replacement of a few roof panels that were damaged by 50 mph winds.

Checking out the latest in barnyard architecture: Hoop barns

A new type of barn is starting to pop up across the countryside: *hoop barns.* These barns are arch-shaped structures with wooden side walls that are 4 to 12 feet high. Tubular steel arches attach to the wooden walls to form a roof. These arches are covered by tarps or heavy plastic.

Here are some benefits of hoop barns:

- **They're affordable.** Materials for hoop barns can be about 20 percent less than supplies for comparably sized pole barns.
- **They're versatile.** The curving side wall of a hoop barn lets you fit more big, round hay bales than you could in a traditional barn with straight walls. If you cover the hoop barn with a translucent plastic, the barns can also double as greenhouses.
- **They're portable.** Hoop barns can be moved to different locations (with varying degrees of effort depending on the size).

Because hoop barns for livestock are a newer type of animal shelter, you may have a difficult time finding a builder with much experience constructing them. Also, we can't vouch for the long-term durability and functionality of hoop barns because they haven't been around long enough yet.

Making do with what you have: Repurposing existing structures

If you don't have the time or the money to build a new shelter for your cattle, you can remodel a variety of structures to serve as suitable shelters. Options include garages, sheds, equipment buildings, and barns originally intended for other species of livestock. Depending on the scope of remodeling and your skills, you can tackle the project yourself or hire someone to do it for you.

Take the time to look over your potential remodeling project to make sure it won't end up costing you more than new construction would. In old buildings, take an especially close look at the following:

- **Foundation walls:** Make sure the most important walls are structurally sound. Repairing cracked or leaning walls can become expensive quickly.
- **Structural frame:** Are the joists, beams, and walls free from termite damage and excessive wear and tear? If not, be prepared for a heftier remodeling bill.

Barnyard advice from the experts

Many organizations provide in-depth resources, plans, and recommendations for designing and building your own handling facilities and shelters, many at little to no cost. Here are a few sources to get you started:

- **MidWest Plan Service** (www.mwps.org): MidWest Plan Service (MWPS) is a university-based publishing cooperative that produces publications for farmers, homeowners, municipal planners, and more. They provide detailed layouts and designs of structures, fences, and handling facilities either for free or for a minimal fee.
- **Your local cooperative extension service:** Cooperative Extension Service personnel come from colleges, universities, and the U.S. Department of Agriculture and can be found across the United States. Most counties have extension service personnel who provide scientific, research-based information and education. To find your local cooperative extension service, go to www.csrees.usda.gov/Extension/USA-text.html.
- **Dr. Temple Grandin** (www.grandin.com): Grandin is an internationally recognized expert in animal handling and welfare. On her website she gives a lot of the "whys" behind what it takes to create an animal-friendly facility.

- **Interior concrete floors:** Are the floors level and well draining? If the floors are sunken, you may need to remove the concrete and build up the floor for better drainage.
- **Electrical and plumbing:** Does the shelter have adequate electrical service and frost-free water systems? Working by flashlight on frozen water lines on a frigid December night is no fun. Having electric and water service connected to your barn is convenient but not absolutely necessary, so if installing water and power is too expensive, you can at least know your cattle are protected from the weather.

Providing shade sources and wind breaks

Hot, humid weather and low wind chills can affect the well-being of your cattle. So you need to have shade sources to cool them off and wind breaks to slow down the wind in the cold winter months. We explain both in the following sections.

Shade sources

Shade sources are critical when temperatures exceed 85° Fahrenheit and humidity levels are high. If you have a well-ventilated barn or shed, you can allow your cattle to seek shelter inside during hot spells. You can provide ventilation by opening doors or windows as well as vents on the peak of the roof. Using fans designed for agricultural use helps increase the air flow and comfort levels of your cattle in a barn during hot spells.

For pasture-based cattle, you can make or purchase portable shades. A *portable shade* is usually made of a 2.5-inch steel pipe frame with an 80 percent shade cloth that cattle can stand under. The frame should have a skid bottom so you can move it from pasture to pasture by pulling it behind your tractor, truck, or ATV. When determining how many shades you need and what size, keep these guidelines in mind:

Type of Cattle	*Minimum Square Feet of Shade*
Calves (≈ 400 lbs.)	15–20 per head
Yearlings (≈ 800 lbs.)	20–25 per head
Cows (≈ 1,000 to 1,200 lbs.)	30–40 per head

You can also use trees to provide shade to pasture-based cattle. When you're selecting new trees to provide shade, consider varieties that are already thriving in your pasture (or in your neighbor's pasture). Match the tree with your area's soil, temperature, and precipitation. Doing so provides better results and is more environmentally sustainable. Also, space the trees throughout the pasture so cattle don't congregate in one area. Ideally, the shade provided by your trees should cover a bit more square footage than what's recommended for portable shades.

Wind breaks

When the weather is cold and windy, you need to protect your cattle with wind breaks. *Wind breaks* are tall walls or rows of trees that slow or divert the movement of air. By slowing the wind, you can reduce the impact of wind chill on your cattle's well-being and on your feed bill.

For a yearling heifer or steer with an average winter coat, every degree drop in temperature below 32° Fahrenheit forces the animal to increase feed consumption by 1.1 percent just to maintain its body temperature.

In barns or feeding areas where trees aren't feasible, you can use a *windbreak fence.* These fences are often made of wooden boards spaced 1 to 2 inches apart with a vertical height of 8 to 10 feet. If both wind and snow are a concern, eliminate the gaps in the fence; otherwise, you'll have problems with wet cattle and snowdrifts.

Where trees are possible, two or three rows of evergreens 60 to 65 feet upwind from the area needing protection can work quite well to block the wind. An effective windbreak should have low permeability, so plant the trees close (every 6 to 10 feet depending on species) and space the rows 5 to 6 feet apart.

Don't plant any trees or shrubs that may be poisonous to livestock. For instance, avoid wild cherry, ornamental yew bushes, ponderosa pines (their needles can cause abortions), and oaks. This list of plants to avoid is by no means comprehensive, and keep in mind that just because a tree or plant is okay for one livestock species doesn't mean it's safe for another. Your best course of action is to plant those trees that are safe for all livestock. Contact your local extension service for toxic plants and trees found in your area. And keep in mind that animals naturally avoid dangerous plants if they have enough good food to eat, so provide plenty of wholesome feed.

Cattle can be rough on trees because they rub on the trees and compact the soil around them. If you have only a few shade trees in your pasture or are starting new trees, use 5-foot cattle panels or wooden boards to fence off an 8-x-8-foot area around the trees. Another way to protect your windbreak trees is to plant them just outside the boundary of your pasture.

Choosing bedding for your cattle shelter

Your cattle need a clean, dry area to rest. Provide this area by using a bedding material that's absorbent and clean. You should use enough bedding to fully cover the rest area and keep the space clean for several days. Don't put down extra thick layers of bedding because the cattle will waste it. It's less wasteful to bed regularly with moderate amounts of bedding than once a week with a whole bunch.

Several bedding materials are suitable for keeping your animals healthier and more content. Depending on where you live, the cost and availability of the different types of bedding can vary quite a bit. Do some comparison shopping to find the best deal for your situation. We explain the options in the following sections.

Straw

Straw is the most widely used bedding material. It's made from the dried stems of wheat, oats, barley, or other small grain plants. You can purchase straw as small rectangular bales, which you can carry and stack by hand, or in larger bales that require the use of a tractor or skid steer to move.

When purchasing straw, make sure it

- **Has been baled dry:** Straw that has been baled without being rained on is cleaner and easier to spread. Straw that's baled wet can be moldy, clumps together, and may be a health concern if your animals consume it.

- **Has little dust:** Dusty bales can irritate the respiratory systems of humans and cattle.
- **Is in tight bales:** Tight bales help the straw hold together better during storage and handling, and they cover more area when opened for use.

Buying straw directly out of a farmer's field is often the cheapest option. You can also buy groups of 25 to over 100 straw bales at livestock sale barns. Feed stores are usually the most expensive source.

Woodchips and corn cobs

You can purchase woodchips and corn cob bedding in bags small enough to be managed by hand or in enough bulk volume to fill up the back of a pickup or semitrailer. Both materials are generally clean and consistent in quality. The corn cob bedding and most woodchips absorb a great deal of moisture, which keeps your cattle dry and comfortable and reduces odors in their resting areas.

These products may be cost effective if high-quality straw is difficult to find or expensive in your area. If you live near a wood mill or other source, bulk woodchips may be available at a lower cost. These sources may also sell sawdust. Sawdust is often dustier than woodchips, but it's another bedding material that's easy to spread, absorbs a lot of moisture, and helps reduce odors.

Newspaper

Chopped or shredded newspaper is dust-free and weed-free, absorbs moisture, and decomposes in the soil. Check with a local or regional recycler to see whether they provide newspaper for livestock bedding. However, keep in mind that chopped or shredded newspaper may not be suitable in areas where it can be blown across fields into yards or other undesirable areas.

Avoid paper bedding from nonnewspaper sources like magazines and colored advertising circulars. Most newspaper print is environmentally friendly, but not all glossy publications are.

Preventing Mud and Muck with a HUAP Zone

For cattle producers in locations that get even a little bit of snow or rain, just a few stomping hooves can turn a pen, barn, or pasture into a muddy mess. And even if you live in an area with little precipitation, you still have to deal with manure buildup. To deal with some of this mess, you can create a *heavy use area protection zone* (or HUAP zone for short).

A HUAP zone is a solid, cleanable surface used in areas where cattle gather, such as around feeders, waterers, barn entrances, handling facilities, or walkways. A HUAP zone protects and stabilizes an area of your farm that gets a lot of traffic from your cattle or equipment, and it can also be used to protect streams and ponds (see Chapter 6 for details).

Your HUAP zone should provide a solid walking surface of concrete (or other man-made material) or crushed rock. Concrete is more expensive than rock, but if the surface gets a lot of manure buildup, it's easier to scrape clean. Also, if the area gets a lot of equipment traffic, solid man-made materials last longer than rock. If you use concrete for your surface material, make sure it's slip-resistant.

To build a HUAP zone with rock, follow these steps:

1. **Determine the areas where cattle congregate, especially in the wet seasons.**

   The areas you'll likely notice cattle congregating include around feeders, waterers, barn entrances, handling facilities, and walkways.

2. **Measure the amount of area you'll cover.**

   For walkways, plan to fix the entire area, not just a strip down the middle. A HUAP zone should extend at least 10 feet from the edges of feeders and waterers.

3. **Calculate the amount of materials you need, including geotextile fabric and crushed rock.**

   Geotextile fabric is a durable yet lightweight cloth material with small openings that water can pass through but rocks can't so they don't get buried in the mud. You can order the fabric from farm supply catalogs; your local aggregate supply company or industrial construction company may also stock it. Plan to overlap each swath of fabric 12 to 18 inches to prevent gaps if the material shifts.

   You need to cover all the geotextile fabric with at least 6 inches of rock. You can use bigger 3- to 4-inch rocks as a base, but your top couple of inches should be rock that's 1 inch or smaller. This size rock provides a smooth walking surface for your cattle. Also, be sure your rock surfaces slope away from the barn, feeder, or waterer to aid in drainage. Keep in mind that if you have to change the slope of an area, you need to plan for more rock.

4. **Clean out all the mud and muck around the area you want to protect.**

   Using a skid steer or tractor with bucket is quicker and easier than shoveling by hand, but either method works.

5. **Place your fabric on the ground, being sure to overlap it 12 to 18 inches. Then cover the fabric with your rock.**

   Take care to not walk or drive on the fabric until it's covered; otherwise, you may rip or move it.

A HUAP zone is a highly recommended conservation practice. Check with your local Soil and Water Conservation District personnel about possible government grants or cost-share programs that may help pay for your HUAP.

Chapter 5

Chow Time: Planning Your Feeding Program

In This Chapter

- Finding out what nutrients cattle need
- Investigating different types of feed
- Creating diets for cattle at a variety of ages and life stages
- Settling on a feeding schedule
- Selecting feeding and watering equipment

Meeting your beef cattle's basic nutrient requirements is critical to maintaining their health and production. The nutrient requirements of cattle change over time based on their age, stage in the production cycle (pregnant, open, lactating, rebreeding), sex, breed, and environment. Growing or purchasing feed is the biggest continual expense when raising cattle, and feeding your livestock is one of most time-consuming daily chores.

Given the importance of proper feeding for adequate animal well-being, growth, and reproduction — and considering that feed and feeding is such a major investment in terms of money and time — it's important to have a basic understanding of cattle nutrition so you can make informed, cost-effective choices regarding your herd's feedstuffs.

We start off this chapter by reviewing the main nutrients cattle need so you can create a balanced diet. Next, we take a look at the main types of feed — forage, grain, and nontraditional feedstuffs — and the pros and cons of these different options.

Tailoring a feeding program to animals ranging from a newborn calf to a 2,000-pound bull takes some adjustments, but we give you guidelines for getting your type of cattle fed right. After you decide what to feed your animals, you must determine how often you should feed and what equipment you need to get the job done; so we wrap up this chapter by discussing feeding schedules and supplies.

Understanding the Nutrients Cattle Need

With so many choices on what you can feed your beef cattle, deciding where to start in developing a ration can be daunting. By first understanding the basic nutrients (protein, energy, minerals, vitamins, and water) that cattle require, you're in a better position to knowledgeably analyze different feed options to determine what's best for your cattle. The following sections provide you with a thorough overview.

Starting with protein

Proteins are complex compounds made of amino acids. The *amino acids* are organic compounds containing nitrogen, carbon, hydrogen, and oxygen. Proteins have many critical roles in the body, including the following:

- They're necessary to build and repair tissue, especially muscle.
- They're involved in several body processes, such as enzyme reactions and muscle contraction.
- They can serve as an energy source.

Animals that don't consume enough protein experience slowed growth or weight loss, poor appetites, lower milk production, and reduced reproductive performance.

Protein is often the most expensive part of the bovine diet, so we suggest you put the protein part of the diet in place first and then build the rest of the needed nutrients around the protein source.

Unlike simple-stomached livestock like pigs or chickens, ruminants like cattle aren't as demanding regarding the quality of their protein. Here's why: Microorganisms in the rumen of cattle can synthesize and provide protein and individual amino acids for themselves and their bovine host. The rumen bacteria break down much of the protein cattle consume to ammonia. The bacteria then use the ammonia to rebuild proteins of varying amino acids. So for your ruminant herds, don't focus just on protein quality, keep protein quantity in mind as well.

Cattle need to be healthy enough and old enough (3 months and older) to have a functioning rumen that's capable of supporting bacteria in order to consume lower-quality protein.

Determining the amount of protein in your feed

Just as your box of breakfast cereal lists the nutritional value of what you're eating, feed suppliers provide nutritional information by imprinting it directly on the feed bag or printing it on a small sheet of paper called a *feed tag*. If your cattle eat pasture or hay, you can find out the nutritional information by sending a small sample of those feedstuffs to a lab for analysis. Your local extension service can help you find feed testing labs in your area (go to `www.csrees.usda.gov/Extension/USA-text.html` to find an extension service near you). The lab will return a feed analysis report detailing the levels of nutrients.

On both feed tags and analysis reports, protein is reported as *crude protein percentage*. This type of reporting is done because determining the amount of nitrogen in a feedstuff is easier and more consistent than trying to figure out the actual proteins. All proteins contain an average nitrogen content of about 16 percent, so the amount of crude protein in a feed is calculated by multiplying the nitrogen content of the feed by 6.25. Because this determination of protein percentage isn't measured directly, it's referred to as the "crude" percentage (CP); and even though it's not a direct measurement, CP is still an accepted way of quantifying protein levels in feed.

The moisture content of cattle feed varies greatly, so when calculating rations or comparing different feedstuffs, always make sure the nutrient values are on a dry matter basis to get the most accurate and consistent information. *Dry matter basis* is a measure of the protein, energy, vitamins, minerals, and fiber remaining in a feedstuff after all the water is removed.

Even with their powerful rumen bacteria, cattle can't extract the same amount of nutritional value from all protein sources. In particular, cattle use nonprotein nitrogen (NPN) with only 25 to 50 percent efficiency compared to other types of protein. NPN is found naturally in feeds like corn silage or alfalfa hay and is also available in processed form from supplements, including urea and ammonium salts. So if you want to use a feedstuff with high amounts of NPN, realize that only a portion of the crude protein listed on the feed tag or feed analysis report may be usable by your animal.

Manufactured urea can be an alternate source of nitrogen in cattle diets, especially when plant proteins like soybean meal are high-priced. However, urea toxicity can occur when cattle consume excess urea. It can cause tremors, rapid breathing, and even death. Take care to follow manufacturer's directions when feeding urea, and introduce urea-containing feeds slowly into the cattle's diet.

Looking at recommended protein intakes

When discussing cattle crude protein requirements in the broadest of ranges, cattle need to eat a diet containing 7 to 15 percent crude protein. Here's a general breakdown of this range to get you started:

- **Newborns and nursing calves:** Youngsters get all the protein they need from their mother's milk. If a mother can't care for a calf, the calf's diet should contain around 15 percent crude protein. Feed calves younger than 3 months a specially designed calf feed if they aren't getting their mother's milk.

- **Cows:** These mamas need from 7 to 11 percent crude protein, depending on their size, milk production, and stage of pregnancy. For most mother cows, feeds like late-spring or early-summer pastures and standing fescue in early winter require no supplementation (for more on pastures see Chapter 6). In addition, legume hay, grass-legume hay, and high-quality grass hay can provide sufficient protein for cows. Later in this chapter, we discuss situations where mother cows may benefit from extra protein.
- **Growing market cattle:** This group of cattle needs 10 to 16 percent crude protein, depending on their current and mature size and desired daily rate of weight gain. For large-framed, growing market animals to reach their full potential, they need either high-quality pastures or additional protein supplements.
- **Bulls:** After these fellows are done growing (usually around 2 years of age), they can do fine on 7 percent crude protein. Younger, growing bulls need around 10 percent crude protein.

Table 5-1 lists various protein supplements and their advantages and disadvantages.

Table 5-1 Protein Supplements

Feedstuff/Percentage of Crude Protein	*Features to Consider*
Soybean meal/48%	Common supplement in the Midwest; may collect moisture if not used quickly
Cottonseed meal/41%	May not be readily available in all parts of the country; comes in several different forms (hull, meal, whole), so be sure to determine the nutritive value of the form you're feeding
Dried distillers grains/27%	A feed byproduct of the ethanol industry; may be difficult to purchase in small quantities
Corn gluten feed/20%	A byproduct of corn processing; can come in a wet or dry form; the wet type needs to be used quickly before molding; watch the mineral content to prevent imbalances of phosphorus, potassium, and sulfur
Pellets/20–36%	A manufactured mix of protein, usually from plant and nonprotein nitrogen (NPN) sources; easy to mix with grain and can be formulated to contain vitamins and minerals; pellets are about the size of a human vitamin pill
Range cubes/20–36%	Similar to pellets, except larger in size; designed to be fed outside on pasture ground
Protein blocks or tubs/20–36%	A manufactured product containing protein from plant and NPN sources; less labor is needed to feed animals with these delivery systems, but some animals may overconsume while others don't eat enough

Even though protein is critical for your cattle's well being, they can get too much of a good thing. Feeding excess protein is not only expensive, but it also leads to higher levels of waste nitrogen and ammonia in the urine and manure. Metabolizing the extra protein can even put stress on the body's organs.

Harnessing the nutritional power of energy: Carbs, fats, and proteins

Energy is the ability to do work. It's not a nutrient, per se, but it's contained *in* nutrients like carbohydrates, fats, and proteins. The energy value of a feedstuff is often expressed as total digestible nutrients (TDN). Providing enough TDN is key to animal health and productivity and, subsequently, to cattle business profitability.

In the following sections, we provide information on the different sources of energy: carbohydrates, fats, and proteins.

Revving up your cattle with carbohydrates

The main source of energy for cattle is *carbohydrates.* These organic compounds are made of carbon, hydrogen, and oxygen. Carbohydrates come in two forms: the easily digestible nonstructural carbs (consisting of simple sugars like glucose or starches containing chains of glucose) and the structural carbs that are digested through fermentation.

Structural carbohydrates are cellulose, hemicellulose, and lignin. Cattle can digest the first two in their rumens, but they can't digest lignin. Lignin is found in increasing amounts in the cell walls of plants as forages mature. So be sure to keep your pastures in a highly digestible, palatable state by maintaining young and growing plants either through controlled grazing or mowing.

Cattle energy requirements vary with the stage of production, size, and expected performance of the animals. The TDN requirements can range from the lower 50 percent to the upper 70 percent. High-quality forages like vegetative legumes and cool-season pastures can provide the bulk of the needed energy in most cattle diets. However, the TDN of lower-quality pastures and hay is often in the 45 to 65 percent zone.

Depending on the quality of your cattle's forage, livestock with high-energy needs, such as growing cattle or lactating cows, may need energy supplementation. Some feedstuffs used for energy supplementation and their TDN are shown in Table 5-2.

Table 5-2 Energy Supplements and Their TDN

Feedstuff	Percentage of TDN
Corn	90%
Whole cottonseed	90%
Dried distillers grains	86%
Soybean meal	83%
Citrus pulp	80%
Oats	76%

Using fats judiciously

Fat can be a source of energy for cattle. Fat sources include cottonseed (19 percent fat), canola (40 percent fat), animal fats (100 percent), and commercial mixes of fats (84–99 percent fat). When you feed fats to your cattle, you mix them with the grain or hay portion or feed them as seeds or in a ground meal. Commercial mixes of animal fats are usually available in a liquid form that lightly coats the rest of the ration.

To prevent digestive upset, keep fat levels at or below 5 percent of the total diet.

Enhancing the bovine diet with plant protein supplements

Feeding plant protein supplements, such as soybean or cottonseed meal, along with low-quality forages can help the energy level of a cow more than feeding grain. Extra protein can boost the digestibility of poor forages and supply energy. Feeds high in starch and sugar, like grain, actually reduce the digestibility of low-quality pasture or hay. In most cases, carbohydrates or fats are better energy sources than protein because they're lower-priced and provide more TDN.

Mighty minerals: Finding the right balance

Beef cattle need *minerals* (solid, inorganic substances) for almost all their metabolic activities. Minerals that are used in larger amounts are classified as *major* or *macro minerals.* The macro minerals are calcium, magnesium, phosphorus, potassium, salt, and sulfur. Those required in smaller amounts are called *micro, minor,* or *trace minerals* and include cobalt, copper, iodine, manganese, zinc, and selenium. We explain all these minerals and how to supplement your cattle with them in the following sections.

If your feedstuffs don't intrinsically supply the required minerals, you can meet your cattle's needs by providing a mineral block, loose granules of minerals, or mixing the minerals into the ration.

Saluting the major minerals

Here's a list of the various major minerals that cattle need to function properly:

- **Calcium (Ca):** Calcium is involved in bone formation and maintenance, blood clotting, and enzyme activities. Cattle that are deficient in calcium can have *rickets* (softening and weakening of the bone), improper bone growth, and poor absorption of nutrients through the digestive system. Cattle need calcium supplements when their diets are predominantly grain-based and low in roughage or if they're eating dry, weathered forage.
- **Magnesium (Mg):** Magnesium is needed for strong bones and teeth and to activate enzymes. A deficiency in this mineral can cause cattle to experience *grass tetany,* which is a sometimes-fatal metabolic disease. This condition occurs most frequently in the spring when cattle are grazing lush forage with high water content because plants in this condition are unable to draw adequate amounts of magnesium from the soil. Symptoms include frequent urination and convulsions, and, if left untreated, it can be fatal.

  Supplemental magnesium is usually given in the form of magnesium oxide, which is also known as Mag-Ox. Unfortunately, Mag-Ox has a bitter taste, and when given as *free-choice* (your herd can eat it when they want, and they can eat as much as they want), cattle may not eat enough of it. To ensure adequate consumption (up to 0.07 pounds per day for a 1,000-pound cow) you may need to mix the free-choice magnesium oxide in a 1:1:1 ratio with salt and corn or dried molasses.
- **Phosphorus (P):** Phosphorus plays many roles in the body. Cattle need it for a healthy skeleton, to help with metabolism of feed, and to serve as a component of DNA (genetic material). Cattle with low-phosphorus diets suffer from poor growth and reproduction and are unable to use feed efficiently. Across the world, cattle need phosphorus supplementation more than any other mineral supplement. Make a supplement available as free-choice at all times to cattle on a pasture or hay-based diet.

  Because calcium and phosphorus work together on the skeletal system, they're usually supplied together at a calcium/phosphorus ratio of 2:1.
- **Potassium (K):** Potassium is required for the proper functioning of many organ systems. Low potassium levels lead to general malaise and poor performance. Grain has low levels of potassium, so you may need to supplement those cattle with diets high in grain. Additional potassium may also be needed while grazing fescue pasture January through March because the potassium levels in this type of grass decline throughout the winter months.

- **Salt (sodium chloride or NaCl):** Salt is needed for pH balance and proper fluid volume in cells. Animals deficient in salt experience abnormal appetites and decreased performance. Provide salt as free-choice — either in the form of a salt block or loose granules — at all times to all classes of cattle.
- **Sulfur (S):** Sulfur is a component of some amino acids, vitamins, and enzymes. Sulfur deficiencies can lower feed consumption and result in a rough hair coat and unthriftiness. Excess sulfur, on the other hand, can interfere with the metabolism of other minerals or cause blindness and convulsions associated with the disease poliocephalomalacia. High levels of sulfur can come from water, molasses, beet pulp, and corn gluten feed.

Remembering the micro minerals

Even though the following micro minerals are needed in trace amounts, they aren't any less important than the major minerals we discuss in the preceding section. Keep the following minerals in mind:

- **Cobalt (Co):** Cobalt is essential to the cattle's rumen bacteria for the production of vitamin B_{12}. Low levels of B_{12} and its component cobalt can lead to loss of appetite, anemia, and muscle wasting. Cobalt supplementation at the rate of 1 ounce per ton of free-choice mineral is recommended during the winter for cows on low-quality forages.
- **Copper (Cu):** Copper is needed for proper blood formation and enzyme function. Signs of deficiency include poor appearance, bleached hair, and anemia. If you suspect copper deficiency, also look at sulfur, molybdenum, and iron in the diet, because these minerals can interfere with copper absorption. Use caution to not feed copper in excess because toxicity can occur. Copper supplementation needs vary by region based on the levels of copper in the soil that plants can take up.
- **Iodine (I):** Iodine is needed for making *thyroxin,* a hormone involved in metabolism. It's also needed for normal development of a fetus. Low levels of iodine result in *goiter,* or enlargement of the thyroid gland. Because nitrates decrease iodine uptake, cattle on high-nitrate feeds should be supplemented with iodized salt.
- **Iron (Fe):** Iron is needed to form the hemoglobin portion of blood. Signs of iron deficiency include reduced appetites and pale mucous membranes. Cattle usually don't need iron supplementation unless they're carrying a heavy load of parasites.
- **Manganese (Mn):** Manganese functions as part of enzyme reactions. Deficiencies have only been noted in some areas of the northwestern United States. These deficiencies are characterized by leg deformities in calves and poor reproductive performance in cows.

- **Selenium (Se):** Selenium deficiency results in *white muscle disease,* which is a condition of muscular degeneration in young calves. Grains and forages raised on sandy or acidic soils may be low in selenium. Selenium deficiencies have been known to occur in Illinois, Michigan, Ohio, and Indiana. A soil test (see Chapter 6) can tell whether the soil on your farm is low in selenium.

  Selenium can be toxic at high levels. However, toxic levels of selenium more often occur in cattle grazing alkaline (high pH) soils in the western United States. Selenium and vitamin E have similar functions, so the amount of one in the diet depends on the amount of the other.

- **Zinc (Zn):** Zinc is needed for the proper functioning of the immune and enzyme systems. Deficiencies are indicated by poor coat, slow wound healing, and overall poor performance. Normal cattle diets contain adequate amounts of zinc.

Supplementing cattle with their minerals

You can supply cattle with their supplemental mineral requirements by mixing all the needed ingredients yourself or by purchasing a commercial premade mix. Unless you're feeding uncommon sources of protein and energy or have a unique growing environment for your grain and forages, a premade mix is often the best choice. Consult with your local vet, extension agent, or feed dealer to discuss whether the growing condition or rations you're using warrant specialized mineral supplementation. (To find contact information for your local extension agent, go to `www.csrees.usda.gov/Extension/USA-text.html`.)

Table 5-3 lists the main minerals that cattle need along with the amounts they require and the maximum levels they tolerate so you can be sure the mineral mix you purchase (or create) matches your herd's needs. The units required are expressed as either parts per million (ppm) or percent dry matter of the total daily feed intake.

For example, if your growing steer is eating 20 pounds of dry matter a day, it needs 0.05 to 0.084 pounds of calcium daily. Here's how you calculate it: 0.0025 calcium percent × 20 pounds dry matter = 0.05, and 0.0042 calcium percent × 20 pounds dry matter = 0.084.

Table 5-3 Mineral Guidelines for Cattle

Mineral Units	*Growing Cattle*	*Pregnant Cow Without Calf*	*Lactating Cow*	*Maximum Tolerable Level*
Calcium, percent	0.25–0.42	0.18–0.23	0.27–0.48	Unknown
Cobalt, ppm	0.10	0.10	0.10	10.00
Copper, ppm	10.00	10.00	10.00	100.00
Iodine, ppm	0.2–0.50	0.2–0.50	0.2–0.50	50.00
Iron, ppm	50.00	50.00	50.00	1,000.00
Magnesium, ppm	0.10	0.12	0.20	0.40
Manganese, ppm	20.00	40.00	40.00	1,000.00
Phosphorus, percent	0.18–0.27	0.18–0.21	0.22–0.31	Phosphorus levels should not exceed calcium levels
Potassium, percent	0.60–0.70	0.60	0.70	3.00
Salt, percent	0.07	0.07	0.10	0.10
Selenium, ppm	0.10	0.10	0.10	2.00
Sulfur, percent	0.15	0.15	0.15	0.40
Zinc, ppm	30.00	30.00	30.00	500.00

Adapted from the National Research Council's Nutrient Requirements of Beef Cattle, 7th revised edition (National Academies Press).

If you have different species of livestock sharing a pasture, be sure the mineral mixture you're using is safe for any type of animal that could access it. Otherwise, you may harm one species while trying to help another. For example, the level of copper appropriate for cattle rations is a toxic level for sheep. You can purchase multispecies commercial mixes that are safe for many different species, but they may not provide the optimum amount of supplement as a species-specific blend may. Your best bet for handling a multispecies situation is to avoid feeding minerals at levels that would be harmful to the most sensitive species and keep a watch for signs of deficiencies.

Cooking up an alphabet soup of vital vitamins

Vitamins are organic compounds needed in small quantities for many different functions and body maintenance. Cattle diets primarily need to be supplemented with vitamins A, D, and E. The rumen bacteria under most conditions can synthesize the needed amounts of the B vitamins and vitamin K.

Because vitamins are critical to animal health, you need to be familiar with their purposes, signs of deficiency, and methods of supplementing. Don't worry. We provide everything you need to know in the following sections.

Vitamin A

Vitamin A plays a role in maintaining the skin and the respiratory, digestive, and reproductive linings of the body. It also helps form bone and is needed for night vision. Cattle with a mild deficiency may eat less and grow more slowly, but they don't have the symptoms associated with more severe cases like night blindness, diarrhea, lameness in the joints, and staggering gait. Cows with vitamin A deficiency may abort or fail to rebreed after calving.

As cattle grow, their need for vitamin A increases along with their size. Here's a good rule of thumb: For every pound of dry matter an animal eats, it needs 1,000 IU (international units) of vitamin A. So if a steer is eating 25 pounds of dry matter a day, he requires 25,000 IU of vitamin A. Mother cows eating a diet with at least 6 pounds of good-quality legume hay don't need extra vitamin A. When on all other diets, pregnant cows need 30,000 IU per day, and lactating cows need 45,000 IU daily.

Vitamin A can be supplied in several different ways:

- You can buy it as part of a free-choice mineral/vitamin mix or as a component of commercial protein supplements.
- You can purchase it as a concentrate to add to made-on-the-farm rations.
- You can inject liquid vitamin A into the muscle for any cattle under stress due to conditions like weaning or transportation. This method of supplementation boosts the vitamin A storage in the liver more efficiently than feeding.

Vitamin D

Vitamin D aids in the absorption and metabolic use of calcium and phosphorus. It's also needed for strong bones and teeth. Cattle lacking vitamin D have poor appetites, digestive upset, and stiffness. Pregnant animals may deliver dead or deformed calves.

Cattle synthesize vitamin D when they're directly exposed to the sun. Animals that are housed indoors and don't eat sun-cured forages daily need 120 IU of vitamin D per pound of dry matter consumed. Vitamin D is usually included in commercially prepared vitamin/mineral mixes.

Vitamin E

Vitamin E serves as an antioxidant. It also works closely with selenium to prevent white muscle disease, which is a condition of muscular degeneration in young calves.

Vitamin E is found in green, leafy forages. If your cattle are eating a high-grain ration with small amounts of roughage or feeds that were dried at high temperatures, supplement vitamin E at the level of 50 IU per day. Vitamin E can be a part of a purchased vitamin/mineral mix.

Vitamin K

Vitamin K is essential in blood-clotting mechanisms. Under most conditions, cattle rumen bacteria make vitamin K in sufficient amounts. In the rare cases that vitamin K levels are low, you can give your cattle Vitamin K orally or as an injection.

Moldy clover hay can cause a vitamin K deficiency, so if your cattle are eating clover hay, check for signs of mold.

Tailoring your feeding program to your cattle

The amount of time, money, and thought you put into feeding your cattle can vary greatly. Generally cattle are resilient creatures, so you don't have to figure their dietary needs to the hundredth of a pound. After reading this chapter, however, you may think otherwise. Don't worry. We simply provide in-depth information so you can be as precise as you want or need to be to reach your cattle-raising goals.

For instance, say you want to graze three or four calves on 10 acres through the spring and summer. If you give them free-choice cattle mineral (they can eat it when they want to and can eat as much as they want to) and fresh, clean water, they'll be good to go. However, if you need your cows to give birth every year in January, you need to be more exacting about meeting their dietary needs. Or, you may be raising cattle in a situation that's somewhere in between in terms of nutrition management. This chapter provides basic, foundational information on feeding cattle; we encourage you to work with your feed dealer, vet, or extension agent to develop a feeding program tailored to your unique situation. (To find contact information for your local extension agent, go to `www.csrees.usda.gov/Extension/USA-text.html`.)

B vitamins

The B vitamins include thiamin, biotin, riboflavin, niacin, pantothenic acid, pyridoxine, folic acid, vitamin B_{12}, and choline. They're involved in the metabolism of other nutrients. After the bovine rumen becomes functional around 8 weeks of age, the bacteria there can synthesize the required level of B vitamins. Younger calves get their B vitamins through their mother's milk. Because cobalt is needed for vitamin B synthesis, check cobalt levels in your mineral mix if your cattle have a vitamin B deficiency.

Water: Providing the ever-important liquid

Cool, high-quality water is essential to the health of your cattle. Your herd needs it for temperature regulation, growth, reproduction, and milk production. Although cattle get some water from their feed, you must make sure they have a constant source of drinking water. When cattle get inadequate amounts of water, they reduce their feed intake, thereby slowing growth. Cows with calves reduce milk production if their water is limited. The following sections provide information on where your water sources should be and how much cattle should drink each day. For more on the ideal water sources and equipment, check out the later section "Ensuring proper water sources."

Considering cattle proximity to water

Ideally, cattle on pasture shouldn't have to travel more than 600 feet to access water. This short distance, which may not always be possible, leads to more efficient pasture use and a more even distribution of nutrients from your cattle's manure onto your pasture (rather than concentrating it around the watering area).

Also, when cattle have only a short distance to travel to water, they go individually, allowing each animal ample opportunity to get a drink. When cattle have to travel a long distance to water, the entire herd usually goes to drink together. In this case, if space is limited, smaller or more timid animals are often pushed out of the way by more dominant cattle. When the dominant cattle leave the watering area, the others often leave with them regardless of whether they have had a chance to drink enough. So if your cattle's water must be at a distance that leads to group watering, make sure you provide adequate tank capacity and space for all your cattle to have an opportunity to drink.

If your cattle usually go to drink as a herd, the water tank you choose should hold at least 25 percent of their daily water requirement and refill in no more than one hour. The water tank should also allow a minimum of 10 percent of the herd to drink at once, allowing 2 feet per head. If water is located so that cattle go to drink individually, the water tank only needs to be big enough for about 5 percent of the herd to drink at once, and the refill time can be as long as four hours.

Determining how much water your cattle need

Water consumption varies with the temperature, diet, age, and production ability of your cattle. Table 5-4 shows the daily water requirements for cattle in gallons per head per day.

Table 5-4 Water Consumption for Cattle (in Gallons per Head per Day)

Outdoor Temperature (°F)	*Requirement for 500-Pound Calf*	*Requirement for 1,100-Pound Lactating Cow*	*Requirement for 1,600-Pound Bull*
40	4.4	8.1	8.7
60	5.5	10.1	10.8
80	7.4	12.4	14.5
90	10.6	17.6	20.6

The ideal drinking water temperature for livestock is 40 to 65 degrees Fahrenheit. Water in this range is cooling during the summer but doesn't reduce body temperature too much during the winter.

Comparing Feeds and Choosing the Best for Your Cattle

After you're familiar with the different nutrients your cattle require (see the earlier section "Understanding the Nutrients Cattle Need"), you're ready to decide what combination of feedstuffs to use. Putting together a ration can be as simple as calling your local feed salesman and doing what he or she recommends or as complex as starting from scratch and making your own special diet. As you research your options, evaluate the feeds according to the following factors:

- **Ability of an animal to digest and use the product:** Make sure the feed is appropriate for the cattle and your goals. Stockpiled fescue, for example, makes great winter feed for cows in the middle of pregnancy but isn't acceptable as a finishing ration for a 1,200-pound steer. (Check out Chapter 6 for info on stockpiling.)

- **Availability:** Use local feeds that are in plentiful supply. For instance, citrus pulp may be an option if you're in Florida, and cornstalks are a good bet for Iowa cattle producers.
- **Cost:** To be sure you're making a fair comparison, look at feeds on a dry matter basis. Rank feedstuffs on the cost per percent of crude protein (CP) or percent of total digestible nutrients (TDN).
- **Ease of handling and storage:** Consider how the feed will be delivered to your farm and how you'll store it. Also keep in mind whether you have the equipment and supplies, like feed bunks and hay rings, to easily feed your cattle in a wholesome manner. If you don't, you may have to purchase these things.
- **Nutritional content:** Choosing a feed that matches your cattle's nutritional needs is important. Make sure the feed provides the balance of nutrients your animals need without the costly excess of another nutrient. (Head to the earlier section "Understanding the Nutrients Cattle Need" for a rundown of what cattle need.)
- **Palatability:** If the feed you provide tastes bad and your cattle won't eat it, it doesn't matter how cheap or nutritious it is.

You have several options for feeding your cattle. You can use forages, concentrates, and nontraditional feedstuffs. The following sections provide information on each.

Feeding forages

Cattle's ability to digest and use forages is one of their best attributes, so capitalize on their power of rumination by feeding as much forage as you can. *Forages* are plants or the parts of plants (except the grain portion) that are eaten by animals. Forages usually contain more than 18 percent fiber. We discuss the two main types of forage, pasture and hay, in the following sections.

Pasture

You can grow pasture grasses and legumes to fit every climate and every season. Chapter 6 goes into detail about designing a forage-feeding program using pasture for at least 9 months and maybe even 12 months of the year.

It's a rare pasture that can supply all the nutrition cattle need throughout the year, so plan to supplement with additional protein and energy as necessary. (The earlier section "Understanding the Nutrients Cattle Need" gives you some ideas for protein and energy supplementation.)

Don't fall into the trap of thinking your pasture is "free." Your pasture needs to produce enough feed to at least equal the value of the payment you would receive if you rented the land out for another use. Chapter 6 helps you estimate how much feed your pasture provides.

Dry forages

Pretty much any type of grass or legume that can be grown for pasture can also be cut, dried, and stored as hay. The nutrient value of hay varies just as pasture does. Even the most dedicated graziers have hay in the barn in case of emergencies like severe drought or ice storms.

Hay is more expensive than pasture because you have to pay for the labor and equipment for harvesting. Big, round or square bales of hay are an efficient way to feed cows if you have the equipment to move and deliver this 800- to 1,000-pound bovine meal.

For a nominal fee, commercial or government labs can analyze your hay for nutrient content. Many local cooperative extension services can loan you a *core sampler* that properly removes samples from the hay bales. If you're counting on hay to be a big contributor to your animal's dietary needs, it's worthwhile to know the exact nutrient profile of the hay.

Adding to the cattle diet with concentrates

Concentrates are feeds that have a high concentration of nutrients. Some concentrates, such as the grains of wheat, barley, oats, rye, corn, and sorghum (milo) are high in energy but have less than 20 percent crude protein. Other concentrates, such as soybean meal, cottonseed meal, and linseed meal, are good sources of protein and energy. All concentrates have less than 18 percent fiber.

Concentrates are often more expensive on a per-pound basis than forages, but they're lower in moisture, meaning that each bite of concentrate has less water and more of the other nutrients. They also take up less storage space than forages. For the most part, concentrates have good palatability, so cattle like the taste of them. In fact, cattle's fondness for concentrates can lead to problems if you allow them to overindulge. If cattle gorge themselves on grain, the bacteria in their stomach can become overwhelmed and unable to properly digest the feed, so it's important to make sure you have cattle's grain feed safely secured so they can't overeat.

Concentrates can help supplement forage in several instances:

- **To reduce the "summer slump":** When cool-season grasses mature in July and August, they can become too low in energy and protein to meet the needs of growing cattle. Adding concentrates to their diet helps them gain weight.
- **To lessen the impact of drought:** Feeding concentrates can help supplement the pasture until rain returns.
- **To help cattle put on fat:** If you're selling your cattle as market animals at a sale barn, they need to have some fat covering their ribs and backbone to fetch a good price. Feeding cattle concentrates is an easier way to get the animals to deposit fat than forage feeding.
- **To keep cows productive:** A cow has high requirements for both protein and energy while she's nursing and preparing to rebreed. Concentrates can be an excellent source of these extra nutrients.

Considering nontraditional feedstuffs

Nontraditional feedstuffs are often the byproducts of food, feed, and industrial processing that may otherwise go to waste. Although these feeds are usually reasonably priced, you must consider more than cost. You need to evaluate other important feed characteristics, including nutritional profile, taste, usability, consistent availability, and ease of handling.

When contemplating using nontraditional feedstuffs, always remember the phrase, "You are what you eat." If you're raising beef, you want to provide the cattle with a wholesome, nutritious diet. Some products may meet your personal standards for appropriate feedstuff, while others may not.

A few of the possibilities that are available include bakery waste, crop/vegetable harvest aftermath, distillers grains, and straw.

Before you feed nontraditional feedstuffs, check with your vet or a livestock nutrition extension specialist to make sure the feed is safe for your animals to consume. If you're in doubt, don't feed it to your animals.

Doling Out the Proper Feed for All Types of Cattle

Feed costs are the biggest ongoing expense in raising cattle, so, unless you want to break the bank, you need to feed your cattle appropriately, not lavishly.

Proper nutrition is necessary for long-term animal health and performance, but it isn't a one-size-fits-all program. In this section, we take a look at diets for cattle of many different sizes and production abilities.

Nourishing the newborn

Feeding the newborn calf can be the easiest or hardest part of raising cattle, depending on whether the mother cow does her job properly. If the mother cow produces enough milk, that milk is the only food a young calf needs for several months. Of course, for the cow to do her work, she needs help from you. We talk more about caring for the mother cow in "Keeping your pregnant or lactating cows fed" later in this chapter.

It's critical that calves receive adequate amounts of *colostrum,* the first milk produced by a cow after giving birth. It has five times as much protein and two to three times as much mineral content as milk made by the cow a few days after calving. Colostrum is also high in fat and energy and is essential for providing calves with antibodies to fight disease (because the calves' own immune systems aren't yet functioning).

A calf should be up and nursing two hours after birth. It needs to consume 2 quarts of colostrum in the first six hours of life and an additional 2 quarts by 12 hours of age. How do you know if a calf is getting enough milk from its mama? If the calf gives a big stretch when it stands up from resting, that's a good sign that it's getting enough to eat.

If, on the other hand, a cow can't take care of her calf properly — she has twins or the calf has an injury or illness that prevents it from nursing — you have your work cut out for you. You will be the sole provider for this calf, so keep frozen colostrum or a commercial colostrum supplement on hand. Here's how:

- You can collect colostrum from cows in your herd that lost their calves for nonhealth reasons.
- You can get a small amount (less than 2 cups) from numerous cows in your herd that have plenty of milk.
- You can purchase the commercial supplement through your vet or feed store. Be sure the replacer hasn't passed its freshness expiration date.

Besides colostrum or milk, you also need a bottle and nipple for the calf to suck, or you can teach it to drink from a bucket. If a calf is very weak, you may need to use an esophageal feeder or nasogastric tube to get the colostrum into the calf's stomach. (For more on care of the newborn calf, see Chapter 12.)

Providing for the growing calf

Calves usually continue nursing until they're 4 to 8 months old. Sometimes, though, calves need extra nutrition before they're *weaned* (no longer nursing). *Creep feeding* is a way of supplementing a calf's diet once it's a few months old but still nursing. A *creep* is an area where feed is provided exclusively for calves. They're allowed access to the creep through a gate with openings big enough for calves to pass through but small enough to keep cows out.

Creep feeding makes sense if you want to increase weaning weights of your calves or if your cow herd's milk production is low. Additionally, calves that are creep fed often handle the transition of weaning better. Cattle with above-average growth potential benefit more from creep feeding than slow-growing animals. Creep feed can consist of grain, limit-fed protein supplements, or high-quality pasture.

After a calf is weaned, it usually goes on one of two different diets: a stocker calf diet or a feeder calf diet. We discuss both in the following sections.

Stocker calves

The stocker calf ration is usually designed to grow the skeleton or frame of the calf along with muscle. It isn't meant to fatten calves. Stockers usually weigh between 300 and 500 pounds at the beginning of this phase, and then move on to their finishing ration at 700 to 900 pounds.

Stocker calves eat a high-forage diet (about 70 percent forage). Depending on the region of the country, the diet can vary. During the winter and early spring, calves in the Texas, Oklahoma, and Kansas regions often graze the green growth of small grains, such as wheat, rye, and oats. During that same time in the southeast, stocker calves feed on Bermudagrass or stockpiled fescue. In colder climates, stocker calves typically graze cool-season annuals in the spring and early summer before moving on to a finishing diet (see the later section "Balancing the diet for a finishing animal").

Feeder calves

The feeder calf ration needs to grow the frame and muscle as well as introduce some fat, so it's higher in energy than the stocker calf diet. As your animal grows in size, you need to proportionately increase the pounds of all the feedstuffs so the animal receives about 3 percent of its body weight in feed daily.

For example, if you want your 600-pound calf to grow 1.5 pounds per day, that animal's ration should include 11.5 pounds of alfalfa hay, 6.5 pounds of pelleted grain (14–15 percent crude protein, fortified with minerals and vitamins), and 0.25 pounds of salt with trace minerals.

Balancing the diet for a finishing animal

Cattle weighing 700 to 900 pounds are ready to eat their finishing diet. A *finishing diet* is what cattle eat in the months before they are butchered. The feedstuffs contained in the finishing diet need to supply protein for continued muscle growth and must have plenty of energy so the animals will put on some fat. A little bit of fat increases the palatability of the beef by giving the meat some juiciness and extra flavor. (Flip to Chapter 13 to see a figure of an animal that's ready to be butchered compared to an animal that needs to put on more weight before it's ready for market.)

A finishing diet should be 10 to 12 percent crude protein with corn making up a large portion of the rest of the diet.

Table 5-5 shows a possible diet for cattle that would enable them to gain 2.75 pounds per day. They can eat up to 30 to 40 pounds of this diet every day.

Table 5-5 Finishing Diet

Feed	*Percent (as fed)*
Corn	72.4
Alfalfa hay	27
Dicalcium phosphate	0.1
Limestone	0.15
Salt, trace minerals	0.35
Total	**100%**

Maintaining the dry cow

A *dry cow* is a female that isn't nursing. In her second trimester of pregnancy, the cow isn't producing milk and the fetus is small, so her nutritional requirements are lower at this stage of production than any other.

A typical diet for a dry cow is 20 to 25 pounds of grass hay a day or 1 to 2 acres of cornstalks plus free-choice hay. If you're experiencing severe winter weather or the cow needs to gain weight, you may need to supplement its feed with higher-quality forage or concentrate (see the earlier section "Adding to the cattle diet with concentrates"). For more on caring for your cattle during severe weather, turn to Chapter 9.

Keeping your pregnant or lactating cows fed

All throughout a cow's life, proper nutrition is important. However, probably no one stage of production is as critical to preventing problems and increasing profits as when the cow is in the third trimester of pregnancy and the time when she's raising a young calf.

If you don't feed your cow properly during these times, you can experience numerous negative consequences, including the following:

- The cow will produce less milk, so you'll have fewer pounds of calf to sell.
- The cow will be slower to start coming into heat after giving birth, so the calving season will be longer and harder to manage the following year.
- The cow's calf will be lighter and weaker at birth, which could lead to more calfhood disease or even death.

Here are two possible examples of daily rations for late gestation: 25 to 30 pounds of mixed grass-legume hay, or 15 pounds of wheat straw with 10 pounds of alfalfa hay.

You can tweak these diets for the lactating cow. Offer free-choice mixed grass-legume hay (instead of limiting it to 25 to 30 pounds) plus 1.5 to 2.5 pounds of grain. Or, if you were feeding wheat straw and alfalfa to your cows before calving, once they give birth you can continue with the same amount of wheat straw but increase the alfalfa to 14 pounds (with this ration you also must supplement your lactating cows with trace minerals, salt, and phosphorus). The other three rations mentioned previously just need salt and trace minerals.

Chapter 11 offers more detail about caring for the pregnant cow, including determining whether she's too fat, too thin, or just right.

Meeting the dietary needs of your bull

A single bull contributes half the genetics of your herd, so he must be fertile and physically fit. To be sure your bull is ready when breeding season arrives, you need to properly manage his nutrition. Bulls need to be in good condition and not excessively fat at the beginning of breeding.

A useful resource from Uncle Sam

Reviewing the huge variety of nutrient needs for various cattle and the nutritional value of so many possible feedstuffs is beyond the scope of this book. A great reference for more in-depth values and requirements is *Nutrient Requirements of Beef Cattle,* 7th revised edition. This book is published by the National Academies Press, which was chartered by the United States Congress to provide information on science, engineering, and health. You can find an electronic copy of this book online at `www.nap.edu/openbook.php?isbn=0309069343`.

Leading up to breeding season, young herd bulls of 12 to 24 months of age need high-quality pasture and around 12 pounds of corn every day. You can substitute 20 pounds of grass-legume hay for pasture, if needed. Your mature herd bull should do okay on high-quality pasture; if he needs to regain weight, add a few pounds of grain. When pasture isn't available, feed your mature bull 30 pounds of good-quality hay (greater than 56 percent total digestible nutrients and greater than 10 percent crude protein). To find out more about selecting and managing a bull, see Chapter 11.

Deciding on a Feeding Schedule

When setting a feeding schedule for your cattle, the most important thing is consistency. To keep your cattle's rumen bacteria functioning properly and to prevent digestive upset, cattle need to be on a routine eating schedule. Cattle can adapt to limit feeding or free-choice feeding, so deciding which way to go may depend more on your schedule than the cattle's. In the following sections, we take a look at the benefits of each method and show you how to introduce new feeds to the diet.

Opting for limit feeding

Limit feeding is when you control the amount of feed an animal consumes. This method of feeding is the opposite of *free-choice* feeding, which is when the animal eats all it wants whenever it wants. If you choose to limit feed, you have to deliver the feed to the animals either once or twice a day. This method requires more work for you, but it also gives you the opportunity to check in on the well-being of your animals on a routine schedule.

Here are the benefits of limit feeding:

- The cattle usually have better feed efficiency. They eat less feed to maintain or grow.
- It can help you reduce feed costs because the animals aren't consuming as much feed.

If you're going to limit feed, you need to be sure plenty of space is available at the feeder so the less dominant animals get their fair share.

You can limit feed the more expensive parts of the ration, such as grain or high-quality hay, while allowing the animals free-choice access to lower-quality forages.

Allowing cattle to free-choice feed

Free-choice feeding is often used to provide growing and finishing cattle constant access to a high-energy diet. Free-choice feeding can reduce labor and increase intake compared to limit feeding, but you need to start and manage it carefully.

When starting cattle on free-choice, begin with a high-roughage ration and gradually transition to a higher-grain diet over about three weeks. By using this gradual transition, the potential for digestive upset associated with high-grain diets can be minimized.

Cattle can also be on a free-choice diet when grazing pasture. Allowing the animals to roam freely and graze whatever they want is a low-labor alternative, but you won't get as much production out of your pasture when it's used this way as compared to limit grazing or rotational grazing (see Chapter 6).

Introducing new feeds to the diet

Cattle don't always make the best choices regarding their eating habits, but you can take steps to keep them out of digestion trouble by introducing new feeds slowly. The following two factors give cattle finicky stomachs:

- **Cattle rarely vomit.** If they overindulge or eat something that disagrees with them, cattle have a difficult time quickly expelling it from the body. Compare this to other animals, such as dogs, which seem to vomit at will when they consume something that doesn't sit well on their stomachs.

- **Rumen bacteria are creatures of habit.** Each type of bacteria has preferred foods to ferment, and cattle build up the type of bacteria that specializes in the type of feed they usually eat. When you introduce a new type of feed, it takes a while for the cattle to develop the bacteria that ferment that particular food. If the diet quickly changes from high fiber to high starch, the bacteria present may not safely digest the new material, leading to a dangerous change in rumen pH called *acidosis*.

To prevent problems, never increase feed more than 2 pounds a day per head on high-concentrate rations or 4 to 6 pounds on high-roughage rations. When changing the composition of the diet, do it in a minimum of three stages, with each stage lasting at least five days before altering the ingredients. And keep in mind that cattle shouldn't have feed intakes that vary more than 4 pounds per head per day.

Purchasing the Basic Supplies: Fine China Not Required

Cattle don't need fancy dinnerware to get the nutrition they need, but they do need feeders and water sources that are durable and easy to use and that minimize waste. Selecting and using the proper feeding supplies helps make sure your cattle stay well-nourished and healthy. In this section, we give you a rundown of all the different supplies you may need.

Getting excited about grain feeders

The type and size of grain feeder you choose depends on the size and number of cattle you're feeding. However, the most common type of grain feeder is a *bunk feeder.* These feeders can be purchased or made by hand, and they can be manufactured from a variety of materials, including wood, metal, plastic, or concrete. Figure 5-1 shows a grain feeder like we use on our farm.

Your feeders need to be sturdy enough to withstand exposure to weather, mud, manure, and the pressure of hungry cattle pushing against them. You should also take into consideration the ease with which the feeder can be cleaned out. You wouldn't want to eat something that's been lying on top of manure, snow, or old feed — and neither do your cattle.

Place feeders in a well-drained area that your cattle can access in all types of weather. If you'll be using the feeder in more than one location, consider the mobility of the feeder. For instance, determine whether it's light enough to load on a trailer or in the bed of a pickup, or whether it's on skids so it can be pulled to a new location.

Figure 5-1: A grain feeder.

No matter what kind of feeder you have, it needs to have sufficient space for all your cattle to eat comfortably. Your bunk feeders specifically need to have the proper throat height, length, and width for the size and number of cattle you're feeding. *Throat height* is the height the cattle need to lift their heads over the rim of the feeder to reach the feed. *Bunk length* and *bunk width* refer to the dimensions of the area where the feed is available for cattle to eat from the feeder. Refer to Table 5-6 to determine appropriate bunk feeder dimensions for your cattle.

Table 5-6 Matching Your Bunk Feeder to Your Cattle

Dimension	*For Calves to 600 lbs.*	*For Feeders 600–1,000 lbs.*	*For Feeders and Cows to 1,300 lbs.*
Throat height (in inches)	14–16	18–21	21–24
Bunk length; limit feeding (in inches/head)	14–22	22–26	24–30

If you plan to have feed available for your cattle all or most of the time (by free-choice feeding), you may consider a *self-feeder* to reduce the frequency with which you have to haul feed to your cattle. Self-feeders consist of a bunk with a feed storage box above it. The feed storage box is open along its length just above the feeding bunk. As cattle eat feed from the bunk, more feed falls into the bunk. Because not all cattle eat at the same time, less bunk length is required for free-choice cattle. Free-choice-fed cattle that weigh less than 600 pounds require approximately 3–4 inches of bunk space per head. For free-choice-fed cattle that weigh 600 pounds or more, provide approximately 4–6 inches of bunk space per head.

Relying on hay rings

A *hay ring* is a strong metal band several feet tall that fits snugly around a large, round bale of hay, as shown in Figure 5-2. It allows cattle access to eat hay while reducing the amount that's wasted from trampling and soiling. Using one or more hay rings is particularly beneficial if you're providing several days' worth of hay at each feeding.

When shopping for hay rings, look for a sturdily built ring that has angled slats for the cattle to reach through when eating. The angled slats reduce the frequency at which the cattle remove their heads from the feeder during feeding, which in turn reduces the amount of hay dropped on the ground and wasted.

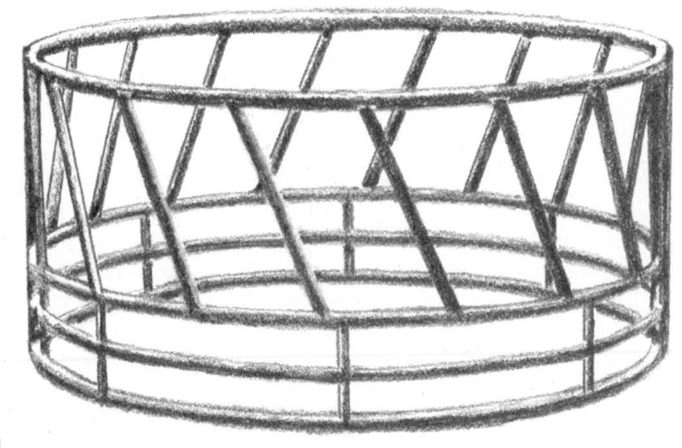

Figure 5-2: A hay ring feeder.

Ensuring proper water sources

Making sure your cattle always have a plentiful supply of clean water is one of the most important things you can do to maintain their health, performance, and profitability. Water can come from any of the sources we describe in the following sections.

A municipal supply or well

Water from a municipal supply or from a well generally provides excellent quality water and allows water to be pumped where needed. You can run pipe from the well to a waterer or use a hose to get water from the well to the water trough.

In order to prevent well-water contamination, don't confine your cattle or spread manure near your well.

A pond, stream, or spring

These natural sources have the potential to be good water sources, but because cattle often loaf in and around bodies of water, you must take care to ensure that they don't contaminate these areas with urine or feces. *Turbidity* (sediment in the water) also increases from hoof action on the bed of the stream or pond, so it's best to limit their access to the source.

To prevent contamination and increased turbidity, fence the water source to exclude cattle or to provide controlled access. Then pipe the water to drinking tanks or allow limited access only where you have built a heavy-use area protection (HUAP) zone. Chapter 6 provides details on ways to protect your streams and ponds.

Tanks or troughs

If your cattle will be drinking from water tanks or troughs, you have many size and material options. A water tank can be as simple as a small plastic tub or as substantial as a large concrete structure. What you use depends on the volume of water and amount of space required for your herd.

Place a *float valve* in your water tank so it automatically refills each time the cattle drink. Float valves can be connected to a garden hose or a quick-connect water line coupler, which is particularly handy if you're moving the water from pasture to pasture in a rotational grazing system. Larger tanks may have the float valve plumbed in with PVC pipe.

Your water tank's float valve should always be located where it can't be damaged or dislodged by the cattle. Also, take care not to leave your float valve attached during freezing weather if its plumbing will be exposed.

If your cattle usually go to drink as a herd, the water tank you choose should hold at least 25 percent of their daily water requirement and refill in no more than one hour. The water tank should also allow a minimum of 10 percent of the herd to drink at once, allowing 2 feet per head.

During freezing weather, you need to take extra steps to ensure that your cattle have a continuous water supply. You can prevent water from freezing by placing your water tanks where exposure to wind and cold air is reduced. You can also use one of the following:

- **Tank heaters:** *Tank heaters* are thermostatically controlled electric or propane-fueled devices that automatically provide heat when the surrounding water temperature falls below a certain level, usually around 35 degrees Fahrenheit. They keep at least a portion of the tank's water thawed so cattle can drink it. Be sure to follow the manufacturer's directions for the safe installation and use of tank heaters.
- **Automatic waterers:** *Automatic waterers* (also called *automatic drinkers*) include electric heaters. Automatic waterers are available in a variety of sizes and styles from many different manufacturers. They're generally constructed of plastic or metal and consist of a trough, float valve, and heater all in one unit. The plumbing and wiring run underground to the waterer, which is set on a concrete base. The initial cost for the installation of an automatic waterer is higher compared to a simple tank and heater, but the convenience factor, particularly in freezing weather, is a big benefit.

When installing an automatic waterer, include a hydrant nearby and a shut-off valve for the waterer. These supplies allow you to stop the water flow to the waterer if it needs to be serviced or repaired, while still having the hydrant as a backup source of water to run to another tank. Figure 5-3 shows a side view of an automatic waterer.

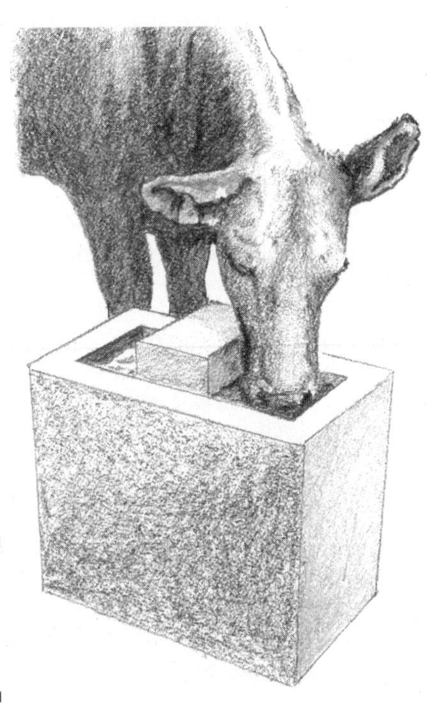

Figure 5-3: An automatic waterer.

Chapter 6

Exploring Pasture, Manure, and Water Management

In This Chapter

- Identifying the components of quality pasture forage
- Starting or improving pasture land
- Balancing pasture production with your cattle's needs
- Protecting grazing areas near ponds and streams

When you think of raising the ideal cattle herd, what do you envision? Is your vision one of plump cows and calves with shiny coats contentedly grazing tall, luscious pastures? Or do you envision skinny, muddy animals picking around at nubs of brown grass? We hope you don't aspire to the latter.

If your goal is to raise healthy cattle in a clean, pleasant environment like the first image we describe, this chapter helps get you started. If your current reality is more like the second scenario, don't despair! With proper planning and plenty of hard work, you can improve the environment and your cattle's well-being.

This chapter discusses proper pasture, manure, and water management. When you implement the tried-and-true techniques we describe, you can increase your pasture's density and diversity of plant life, improve the organic matter and fertility of the soil, and help keep your farm's water cleaner. All these factors translate into healthier cattle, more money in your wallet, and an altogether more enjoyable cattle-raising experience.

Pasture Basics: The Forage Elements that Make for Good Grazing

Cattle are well-equipped to gather their food by grazing. Their four-compartment stomachs enable them to digest tough, fibrous plant materials that would be unpalatable and useless to creatures — such as humans — with simple stomachs. (For more on cattle stomachs, see Chapter 2.) The more you can capitalize on your cattle's grazing ability, the healthier and more naturally raised your cattle will be. Plus, by providing cattle the opportunity to feed themselves, you use less of your energy and money harvesting their food and feeding them.

In order to allow cattle to sustain themselves by grazing, you must have the proper forages available for them. So in this section, we explain what makes a high-quality forage and introduce some great varieties to try in your pasture.

Characterizing quality forages

Your goal should be to maintain a pasture that produces high-quality forages. *Forages* are plants or the parts of plants (except for the grain portion) eaten by livestock.

Quality forage has these characteristics:

- **Low in fiber:** Although the design of your cattle's digestive system enables them to use the fiber found in plants, they have a limit to how much nutritional value they can derive from extremely fibrous plants.

  The key to keeping fiber levels down is to prevent plants from maturing. Cattle do this themselves by grazing the pasture, or you can do it by mowing. Trimming forages by cow or mower signals the plants to add more vegetative, young growth; if they aren't trimmed, the plants continue to mature, increasing their fiber content and forming seeds.

- **High in protein:** Protein is the building block of muscle, and forages high in protein provide more of this nutrient that your animals need for growth.

  As a plant matures, the amount of fiber increases and the amount of protein decreases. To help your cattle get adequate levels of protein, have them graze young, growing plants when possible.

- **Tasty:** Just like any animal, cattle prefer to eat things that taste good to them. Cattle avoid plants that produce chemical compounds that give them a bad taste or cause them to become sick. (These plants probably create these bad-tasting compounds as a mechanism to protect themselves against being eaten or against disease.)

Avoid or at least minimize unpalatable plants in your pastures. Don't seed these plants in your pasture, and take into account the presence of these plants when figuring the value of any pasture you're considering renting or buying. Some common examples of forages that taste bad or that are detrimental to cattle include endophyte-infected fescue (*endophyte* is a fungus that lives in the fescue plant; cattle that eat lots of endophyte-infected fescue can have lower growth and reproductive performance), sericea lespedeza (the original, unimproved variety; the modern varieties can be useful forages), and sagebrush containing *terpene* (a strong-smelling chemical produced by some plants).

Looking at the types of forages that are available for your pastures

Before you can improve — or start — a pasture for grazing, you need to become familiar with some of the basic types of forages. The two main types of forages used for pastures are legumes and grasses. We explore these and a few other common types of pasture forages in the following sections. (See Chapter 5 for details on non-pasture forages.)

Lovin' legumes

Legumes are plants that are a good source of protein for your cattle. We list some common legumes, along with their benefits and drawbacks, in Table 6-1.

Legumes work with bacteria on their roots to capture nitrogen from the air. This process is known as *nitrogen fixation.* The plant relies on nitrogen for growth, and the plant also increases soil fertility by depositing excess nitrogen into the soil so other plants, such as grasses, can use it. The more nitrogen legumes put in the soil, the less fertilizer you need to add for proper grass growth.

Table 6-1 Legumes for Your Cattle Pasture

Legume	*Benefits*	*Drawbacks*
Alfalfa	Highly digestible; high rate of nitrogen fixation; drought resistant	Needs pastures with good drainage; certain varieties are susceptible to insects and diseases; may not hold up to heavy grazing
White clover	Holds up well during grazing	Has a shallow root system that requires plenty of moisture

(continued)

Table 6-1 *(continued)*

Legume	*Benefits*	*Drawbacks*
Red clover	Tolerates poor drainage and acidic soils	Has poor drought tolerance; short plant life often requires it to be replanted every few years
Sericea lespedeza (use only improved varieties that are more appealing to cattle than the older varieties)	Drought resistant; tolerates acidic soils	Is an annual plant that must be reseeded every year

Be careful of too much of a good thing. Even though alfalfa and clover make great cattle feed, if cattle overindulge on either of these two legumes, they can suffer from a life-threatening condition called *bloat.* For more about preventing and treating bloat, see Chapter 10.

Growing grasses

With more than 9,000 species worldwide, grasses are a common and important forage. You can find varieties that thrive in most any grazing situation and provide cattle with energy and some protein. Grasses are characterized by their long, slender leaves that usually grow in an upright fashion. Grasses are good at using the nitrogen found in the soil to fuel their growth. Table 6-2 describes some popular forage grasses.

Table 6-2 Grasses for Your Cattle Pasture

Grass	*Benefits*	*Drawbacks*
Fescue	Hardy and long-lived; good for winter grazing	Can be infected with a type of fungus (endophyte), and consuming high levels of infected fescue decreases animal performance and can cause "fescue foot"
Kentucky bluegrass	Long-lived; high palatability	Unproductive during hot weather

Grass	Benefits	Drawbacks
Smooth bromegrass	High in protein; very palatable	Newly planted brome-grass pastures don't tolerate heavy grazing; slow growth in the summer
Orchardgrass	Drought tolerant; high palatability	May be difficult to establish; doesn't tolerate being grazed short
Perennial ryegrass	High quality; tolerates heavy grazing	Many varieties not cold hardy
Bermudagrass	Good growth during hot weather	Won't survive through cold winters

Other common types of pasture forage

Pasture forages aren't limited to legumes and grasses. They also include the following:

- Leaves and twigs of trees and shrubs
- Broadleaf plants, such as dandelion and chicory
- Crops like corn, wheat, and sorghum (as long as the plant hasn't developed seeds or grain)
- Members of the brassica family of plants, including turnips, kale, and forage rape
- *Crop aftermath,* which is the plant material remaining after grain harvest

You can use these other forages to supply nutrients and energy when your traditional legumes and grasses aren't at peak production or simply aren't available. However, keep in mind that trees, shrubs, and most broadleaves produce less feed per acre than grasses and legumes. Unlike many of the legumes and grasses that are *perennials* (plants that grow back ever year on their own), you must sow crops and brassica plants annually.

Seeking Greener Pastures: Improving or Establishing Your Grazing Land

Whether you have fields that have been pasture for as long as you can remember or you're starting with bare soil, you can transform your cattle grazing land into quality pastures. The following sections provide you with the tools you need to improve the quality of existing pastures or to start a brand new pasture from scratch. We also provide information on amending your soil and managing manure, two important topics when tending to your land.

Evaluating and perfecting an existing pasture

If you currently have pasture, plan to take several pasture walks throughout the year to gauge its quality. A *pasture walk* is a regular, systematic observation and evaluation of your pastures; it's more than just a stroll across your field. For a good, random sample of your entire pasture, walk across the land in a W-shaped pattern, stopping to take assessments at 10 to 20 different spots. Take pen, paper, and a ruler so you can record the following data:

- **Types and percentages of forages you see for each of the seasons:** For better nutritional value and increased months of productivity, you should have different forages (both grass and legume) throughout the year. You can determine percentage by counting the different types of plants in a 2-feet-square area as you take your other measurements. Pasture seed catalogs often have great descriptions and pictures of different forages if you need help identifying types of forage.
- **Height of the pasture measuring from the soil to the top of the plant canopy:** The *plant canopy* is the level where, when you place your hand down, the plants are thick enough to give resistance to your touch; it isn't the tallest stem of the tallest plant. (We talk more about the importance of foliage height in the later section "Estimating your pasture yields.")
- **Density of the forage:** Look down where you're standing to determine the amount of ground covered by forage that's tall enough for cattle to eat (usually 6 inches for most types of forage). Plant density is usually classified as follows:
  - **Poor:** Less than 75 percent coverage
  - **Good:** 75 to 90 percent density
  - **Excellent:** More than 90 percent of the ground covered
- **Maturity stage of the plants:** Are your plants green, young, and growing (vegetative); or have they developed seeds, stopped growing, and become coarse (mature)?

If, throughout the seasons, your pasture walks reveal a mix of legumes and grasses covering more than 90 percent of the ground at a height of 6 to 14 inches, keep up the excellent work! If your pastures don't quite hit this ideal, you're not alone. However, you can take steps to make improvements. Table 6-3 reviews some pasture challenges and possible solutions.

Table 6-3 Pasture Problems and Solutions

Problem	*Solution*
All grasses, no legumes	Spread clover or newer varieties of sericea lespedeza seed over your pastures between late February and early March.
Plant canopy less than 4 inches	Reduce the number of animals grazing on the pasture, or increase the time for pasture recovery between grazing periods.
Poor plant density	Reduce the number of animals grazing on the pasture, or increase the time for pasture recovery between grazing periods. Also, you may need to add fertilizer and nutrients to the soil (more on this later in the chapter).
Many mature plants	Increase the number of grazing animals or graze more frequently, and be sure to diversify with plants of different maturity rates.
Excess pasture at some times and not enough at other times	Change your production cycle to better match your forage supply, or switch some fields to new forages to move production to when it is needed (more on this later in the chapter).

Preparing a new pasture

If you have nothing but a plot of dirt, don't worry. When starting from scratch, you have the opportunity to prepare your soil and make the best match among your land, climate, and new plants. Before you plant your first seed, follow these steps to encourage a productive pasture:

1. **Take a soil test to determine the pH and figure out what nutrients, if any, you need to add.**

   The pH of soil is important because plants need the appropriate range for proper growth. You can read more about soil samples in the later section "Improving your pasture by fertilizing and feeding the soil." Doing a soil test can be beneficial for established pastures, too. A soil test can tell you what nutrients are present in the soil and what nutrients, if any, you need to add.

2. **Determine the soil type of the land you're working on.**

   Soil can be clay-like, sandy, or loamy. Soil that's clay-like holds water and leads to wet conditions. Sandy soil dries out quickly and isn't the best home for plants with high water needs. Loamy soil is good for most pasture forage.

Your local Natural Resource Conservation Service (do an Internet search for "[your state] NRCS" to find an office or personnel close to you) or county conservation district staff (search for "[your state] SWCD") can supply information on soil types.

3. **Match your plant characteristics to your soil characteristics (pH and type), and match the plant requirements to your local temperature and precipitation levels.**

   By selecting forages that are best suited for your ground and environment, you have the best chance of growing a vibrant, long-lived pasture.

The wide array of plant options and climates across the country make it impossible to give you a one-size-fits-all pasture planting guide. One of the best ways to find the right mix is to look at forages that do well in your area. You can touch base with the following groups to determine what works for your area:

- **Local cattle producers:** Talk with other cattle producers in your area to find out what forages they like and what they would change about their pastures.
- **County extension service:** Extension office personnel can offer lots of production and adaptability data for forages in your area. You can locate an extension office near you at `www.csrees.usda.gov/Extension/USA-text.html`.
- **Your local seed salesperson:** Although their primary job is to sell product, seed company reps can be a wealth of information, especially when it comes to local planting and fertilizing advice.

Improving your pasture by fertilizing and feeding the soil

Your cattle are healthier if they have ample, nutritious forages to eat, but it's hard to grow good forages if the soil in your pasture has low fertility and lacks the nutrients plants need for proper growth. You can often increase the production of pastures by fertilizing with nitrogen and adding nutrients such as phosphorus and potassium. By the same token, your pastures need proper soil pH for the best forage production.

Before you can make informed decisions about what nutrients (if any) to add to your pastures, you need to find out their current nutrient levels. To get that information, you can take a soil sample and then investigate its results. Consult with your local extension service for advice on how to best take soil samples and where to find a lab to perform the analysis.

The four main soil test results focus on soil pH and the *macronutrients* (nutrients needed in larger amounts, as opposed to *micronutrients* that are needed in trace amounts, if at all) of nitrogen, phosphorus, and potassium. Every pasture has its own unique needs, so we can't give you specific remedies. However, the following information can help you make informed decisions when you work with your fertilizer supplier and seed salesman to pinpoint the needs of your particular fields. Here's a rundown of the four categories:

- **pH:** Soil pH indicates the acidity or alkalinity of soil based on a scale of 0 to 14. A pH of 7.0 is neutral; values below that are acidic, and values above that are alkaline. Different forages have preferred pHs, but most do fine in the 6.5 to 7 pH range.

  To raise soil pH, apply lime. Lowering pH for a large area is difficult, so you may need to compensate by increasing the levels of macronutrients.

- **Nitrogen (N):** This element is most often the limiting macronutrient impacting plant growth. It can be applied in a variety of forms, including urea, ammonium nitrate, or organic solids like manure.

  The nitrogen needs of forages can vary greatly. Summer annuals require more nitrogen than perennial grasses, and legumes traditionally need the least. Nitrogen fertilization may increase grass yields two to three times compared to unfertilized fields. You can boost pasture growth and stockpiling potential by applying nitrogen fertilizer once or twice a year. If you have plenty of spring pasture, you may opt to only fertilize your cool season perennials in late summer to boost fall growth. If you need more spring and fall forage, fertilize in April and again in August.

  Don't waste your money fertilizing in August if you're in the midst of a drought. Most of the nitrogen will be lost to evaporation before the plants can use it.

- **Phosphorus (P) and potassium (K):** Phosphorus is essential for the overall health and vigor of plants. Potassium is needed in nearly all plant growth and reproduction processes.

  Phosphorus and potassium are returned to the soil in cattle urine and manure, so you usually need lower levels of these elements than nitrogen when fertilizing your field. The exception is pastures that primarily consist of legumes, because phosphorus and potassium are important to legume plant health and yield.

Making manure a nutrient for your pasture, not a nuisance

If your cattle spend plenty of time out on pasture, they'll distribute most of their manure back on the land without any help from you. However, manure still has a way of accumulating in barns and around feeders and waterers.

This manure contains many nutrients that can improve your pasture's soil, but to get the most benefit and not cause pollution, you need to take care to apply it properly.

In this section, we give the reasons and steps for composting, plus we offer guidelines for safely and effectively spreading the manure or compost on your fields.

Making compost

Composting is a managed process in which you help control and monitor the breakdown of plant and animal materials by microorganisms into more usable materials that are suitable for application to the soil.

Using compost instead of raw manure to fertilize your pastures has the following benefits:

- **Safer:** A well-managed compost pile can reach temperatures higher than 130 degrees Fahrenheit. These high temperatures kill fly eggs and many pathogens, thereby reducing pest and disease problems. High temps also cause weed seeds to lose the ability to germinate.
- **Less smelly:** Properly maintained compost piles are less smelly than raw manure.
- **Provides stable nitrogen formation:** Unlike the nitrogen in fresh manure, which is more likely to evaporate or wash away with rain, 95 to 97 percent of nitrogen in compost is in a stable form that can be used by plants over a longer period of time. This more stable, less soluble form of nitrogen isn't as likely to contaminate surface and ground water.

Follow these steps to build a compost pile:

1. **Choose a location for your pile.**

   This location should be far (at least 200 feet) from your home or business and water sources, such as wells, ponds, and streams. Also avoid areas where water collects. When possible, put the compost pile on a solid surface, such as a HUAP zone (short for *heavy use area protection zone;* see Chapter 4 for details) or concrete pad, and cover it with a tarp to minimize waste runoff.

2. **Spread layers of manure and other natural materials, such as leaves, sawdust, grass clippings, and garden or kitchen waste at the location you chose.**

   To accelerate the composting process, it helps to layer the different materials according to their colors. The brown layers, which are high in carbon, include dry leaves, sawdust, and straw. Green materials, which are high in nitrogen, include manure, garden or kitchen waste, and grass clippings. Start with a brown layer and then alternate green and brown. Aim for a pile that's three parts green material to two parts brown material (you can achieve this by making your brown layers thinner than the green layers).

Keep an eye on the size of your pile. The pile needs to have enough mass to hold heat yet be small enough so air can reach the center. For large amounts, a compost row 5 feet tall by 5 feet wide by any length is sufficient.

3. **Maintain your pile to speed up the composting process.**

   Turn your pile once a week, either by hand using a shovel or pitchfork, or mechanically with a bucket loader on a tractor. Turning your pile regularly helps you yield compost in a few months. To speed the composting further, water the pile to keep it moist but not wet. Without turning your pile or adding water, the composting process takes anywhere from six months to two years.

When your compost pile is no more than 10 degrees warmer than the air temperature and is a third of its original size, the composting process is complete.

Applying manure and compost properly

The best time of year to spread manure or compost is when you have growing plants that can quickly absorb the nutrients and increase forage quantity and quality. Depending on your pasture forages, you may spread in the spring, summer, or fall.

Unless you're in a warm-weather climate where forages grow year-round, avoid applying manure or compost in the winter, because there won't be any growing plants to use it, and the manure will be more prone to sit on the frozen ground until it gets washed away with snows or spring rains. Also avoid spreading manure on wet ground. Driving on wet soil causes compaction, which decreases the productivity of the soil.

The most common way to spread manure or compost is with a manure spreader. A *manure spreader* is a big box on wheels with a conveyor chain that moves the manure to the back of the implement. At the rear of the spreader are multiple metal plates and spikes that break up the clumps of manure and fling them into the air before they fall to the ground.

Manure spreaders can either be powered by a tractor or ground-driven. *Ground-driven spreaders* can be pulled by a truck or even horses; the conveyor chains and metal plates of the spreader are powered by the rotation of the spreader wheels going over the ground.

Take care not to spread manure too close to homes, roads, and waterways. Table 6-4 shows the recommended distances between manure application areas and structures or waterways.

Table 6-4 Recommended Distance for Manure Spreading

Physical Feature	*Distance*
Streams, rivers, and ponds	100–300 feet (the steeper the incline of the land, the greater the distance required)
Individual homes or businesses	100 feet
Housing developments or non-farm businesses	300 feet
Public roads	25 feet

A ton of fresh manure supplies approximately 10 to 20 pounds of nitrogen per acre, 5 to 10 pounds of phosphorus, and 10 to 15 pounds of potassium. Up to 25 percent of the nitrogen evaporates within 24 hours of spreading manure on the soil surface. Use the data supplied by your soil test to determine how thickly to spread your manure to achieve the proper level of the nutrients in your soil.

Optimizing Pasture Productivity

Managing your pasture to achieve optimal productivity helps give your cattle the nutrients and clean environment they need. A well-tended pasture yields more feed not only for your livestock but also for the wildlife that resides in your pastures. Vibrant pastures can also absorb larger amounts of rainfall than low-density pastures, thereby reducing erosion and water contamination.

In this section, we show you how to achieve that optimal productivity by matching the needs of your cattle to the available forage supply. We also help you increase your forage quality and quantity by using rotational grazing. Finally, we explain how to extend the grazing season for your cattle by creating a forage chain in which you provide a succession of different forages in different pastures based on the season.

The more species the merrier

As more and more of our landscape becomes developed for buildings and roads or used for a single type of crop, many species of wildlife are losing their habitats. Well-managed pastures can provide food and shelter for wild animals as well as cattle. When you see bees, butterflies, songbirds, snakes, frogs, fish, rabbits, fox, deer, and other animals in your cattle pastures, take pride in knowing you're providing a much-needed healthy environment for many living creatures.

Matching the number of cattle to the amount of available pasture

Your *stocking rate* is the number of animal units per unit of land area. This rate has a big impact on animal performance and overall profitability. Getting your stocking rate right is sort of like the story of Goldilocks and the three bears. You need the perfect balance; otherwise, you may overstock or understock:

- *Overstocking* (too many animals in an area) causes desirable forages to be overgrazed and reduced in number while less desirable forages increase.
- *Understocking* (too few animals per acre) results in *patch grazing* (or *spot grazing*), meaning that cattle return again and again to the same area to eat young, tender plant growth while the rest of the pasture becomes mature, less palatable, and less nutritious.

The key to achieving a proper stocking rate and, consequently, matching your cattle to your pasture is to balance the performance of each animal and the vitality of your pasture.

Calculating how much your cattle eat

To determine your stocking rate and, ultimately, the amount of pasture you need, you first have to figure out how much your cattle eat. Unlike human food, which is measured in calories, livestock feed — including pasture, grain, and hay — is usually quantified on a *dry matter basis* (the part of feed that remains after all water is removed). To learn more about the importance of dry matter, see Chapter 5. Table 6-5 shows the amount of dry matter various classes of cattle need.

Table 6-5 Dry Matter Requirements for Cattle

Cattle Type	*Pounds of Dry Matter per Day*
300-pound calf	9
400-pound calf	12
500-pound calf	15
600-pound calf	18
1,000-pound cow, no calf	20
1,000-pound cow, with calf	26
1,200-pound cow, with calf	29
Mature bull	32

Based on Table 6-5, if you have a herd of ten 1,200-pound cows with calves, they would eat 290 pounds of feed per day, because each cow requires 29 pounds of feed per day. When you multiply 29 times 10, you get 290 pounds.

If you don't want to memorize the entire table, the rule of thumb is that growing cattle need to consume approximately 3 percent of their body weight in feed (dry matter basis) each day. Mature animals need around 2.6 percent of their body weight.

Estimating your pasture yields

To determine how much forage your pasture yields, you need to estimate the *dry matter production* (the amount of forage produced with all water removed) in pounds per acre for every inch of forage height. Evaluating pasture production (and, for that matter, all cattle feeds) in terms of dry matter eliminates the variability in weight and volume of feeds due to water. Table 6-6 summarizes the dry matter production for various types of forage. The numbers in the table assume that your pasture has good plant density (75 to 90 percent coverage; see the earlier section "Evaluating and perfecting an existing pasture" for more on this coverage).

Table 6-6 Dry Matter Production for Various Types of Pasture

Type of Pasture	*Pounds of Dry Matter per Inch per Acre*
Tall fescue fertilized with nitrogen	350–450
Tall fescue and legume mix	300–400
Bluegrass and white clover	300–400
Mixed pasture	250–350

For example, if your acre of tall fescue and legume mix pasture is 10 inches high and you want to graze it down to 4 inches, you'd have 6 inches of available forage. Table 6-6 shows that this type of pasture can produce anywhere from 300 to 400 pounds for every inch of height per acre. Your 6 inches of available forage would yield 1,800 to 2,400 pounds of forage. In the preceding section, we calculated that ten 1,200-pound cows and their calves need 290 pounds of feed daily. If your pasture contains 1,800 to 2,400 pounds of pasture production, it would feed your cattle for 6 to 8 days (1,800 ÷ 290 = 6.2 days; 2,400 ÷ 290 = 8.2).

Giving your pastures a rest with rotational grazing

One of the great things about pastures is that they're renewable resources. If you give your pastures time to recover from grazing, the same acre can feed your herd more than once every year. This concept of grazing pasture and then letting it rest and grow back is called *rotational grazing.*

Rotational grazing is beneficial in the following ways:

- **It keeps plants from maturing,** so they're more digestible and nutritious (for details, see the earlier section "Characterizing quality forages").
- **It encourages plant diversity** because cattle don't remain in a paddock long enough to eat one particular type of forage in excess. In times of plentiful pasture, you can mow some paddocks for hay.
- **It helps decrease internal parasites and flies** because when the cattle move to a new paddock, they leave some of these pests behind.
- **It increases forage use.** In continuously grazed pastures, where the same set of cattle live on the same pasture year-round, the cattle only use about 30 to 40 percent of the pasture. For systems where you move cattle once or twice a day, less than 15 percent of the pasture is wasted. Rotational grazing results in less waste because cattle have less pasture to choose from and can't be so picky, ignoring forages that aren't their favorite.

To follow the practice of rotational grazing, you need to subdivide a pasture into several smaller paddocks with a pasture height of at least 6 inches and preferably 10 to 12 inches. The cattle graze the different paddocks in a set sequence: When the cattle eat the pasture height of their current paddock down to 3 to 4 inches, you move them to a new paddock.

An excellent fence for subdividing pasture is the electric polywire. These fences are relatively inexpensive and quick to install and take down. Figure 6-1 shows how to effectively use a single polywire fence to encourage animals to eat their pasture fully. (Refer to Chapter 4 for more on fencing.)

The number of paddocks you need to fence depends on the number of days the animals graze a paddock and the length of the pasture rest period. Based on our calculations for a ten-head herd (see the earlier section "Estimating your pasture yields"), an acre provides enough feed for six to eight days. However, if you subdivide that acre into three sections, you can increase the days of feed provided to at least nine days with the cattle on each paddock for three days.

Figure 6-1: Subdividing pasture for rotational grazing using poly-wire fence.

Rest periods for rotationally grazed paddocks can be as short as 10 to 14 days when pastures are growing quickly under optimal conditions, or as long as 40 to 60 days when growing conditions are poor due to heat or drought.

Here's how you figure the number of paddocks you need:

Days of rest per paddock ÷ Days of grazing + 1 extra paddock

So using our earlier example, say you need 30 days of rest after a three-day grazing period. Here's what you get:

$30 \div 3 + 1 = 11$

You need 11 paddocks for this system. With 3 paddocks per acre, you need a little less than 4 acres to support your ten cows and calves through the growing season.

Creating a yearlong forage chain

One pasture probably can't feed your cattle 365 days a year. So you need to make the most of your available land. To help your pastures provide feed for your cattle (instead of having to haul bags or bales of feed), you can create a chain, or succession, of different forages in different pastures to provide feed for many (or all) months of the year. The following sections show you what to grow in each of the calendar months in order to create an effective chain.

January, February, and March

Stockpiled forage is often the best (and only) choice during January, February, and March. A *stockpiled forage* is a forage that's allowed to grow several months without being grazed. You leave it unharvested and standing in the field until the cattle graze it during the winter months when the plant is dormant and not growing. Although some stockpiled forages aren't as young and tender as fresh spring pastures, these stockpiled plants are an excellent way to nourish your cattle compared to the time and expense of buying and feeding hay.

Fescue makes excellent stockpiled feed. It actually becomes more palatable with time and cooler temperatures, and it maintains its shape and structure through wind, rain, and snow. In southern climates, grazing fields that have wheat plants growing in them is an option, and you can also use stockpiled bermudagrass.

April, May, and June

Spring and early summer are often the easiest times of year to pasture feed cattle because *cool-season forages,* such as fescue, perennial ryegrass, bluegrass, bromegrass, and clover, grow rapidly during these months. Pasture comprised of these forages should be more than sufficient for your herd during this time.

Don't be tempted to turn your cattle out onto your established pastures too early in the spring. Let the pastures reach 6 to 8 inches before grazing starts. Giving your forages this head start helps the pasture become more vigorous and productive for the entire growing season.

July, August, and September

Because of the heat and often dry conditions associated with the summer months of July, August, and September, the growth of cool-season forages slows considerably. Warm-season annuals work well during these months. *Warm-season annuals* are plants that thrive in warm or hot weather and need to be planted every year. These plants include sorghum-Sudan grass, grazing corn, and modern varieties of sericea lespedeza. In northern climates, cool-season forages start to make a resurgence in September.

Don't graze sorghum-Sudan grass for at least two weeks after it rains if the plants have suffered through drought conditions. Also, wait at least two weeks after frost before allowing animals to graze sorghum-Sudan. In these two instances, high levels of the toxin prussic acid can accumulate in the plants, and livestock that eat the sorghum-Sudan grass can become sick or even die.

October, November, and December

Cool-season forages continue to grow, albeit slowly, throughout October, November, and December, until they're hit by a killing frost or below-freezing temperatures. They can be supplemented by grazing wheat that you planted in August or early September. Forage brassicas like turnips are ready to eat in October and may last into January. After harvest, crop aftermath from corn or even vegetable crops, such as pumpkins and broccoli, also makes good feed.

Protecting Your Water Resources

Water is important on your farm, right? Of course. So that means you must take care of its sources and the environment around those sources. These areas are called *riparian areas.* They include the streams, ditches, creeks, rivers, ponds, and lakes and all the soil and plants along them. Water quality is greatly influenced by the use and management of these areas.

In this section, we discuss methods of protecting streams and ponds by limiting cattle access to these areas. Many riparian areas offer abundant forage that you want to take advantage of without damaging fragile shorelines. We explain how to use these sensitive areas in a process called *flash grazing.*

Building a stream crossing and drinking area

A stream crossing (see Figure 6-2) provides a stable area where your cattle can cross a stream and get a drink without damaging the streambed or banks. With a stream crossing, the water stays cleaner because the cattle won't loaf in the stream, and the cattle stay healthier by staying out of the mud.

Always check with local authorities to obtain any necessary permits before working in streams. You may also need permits from state authorities and the U.S. Army Corps of Engineers.

When building a stream crossing, you need to make the stream banks flat enough so the cattle can move safely up and down them. To stabilize the banks and stream beds, lay down a layer of geotextile fabric and cover it with large rocks (make sure they're big enough that they won't wash away with rushing water). The rocks provide a stable base for the cattle to walk on but are uncomfortable enough that the cattle won't tarry in the stream.

Figure 6-2: Protecting your riparian areas with a stream crossing.

Install fencing along either side of the crossing and along the stream to make the cattle cross only where you want them to — not where they think they should. However, keep in mind that some streams are difficult to fence because they overflow their banks with great force during times of heavy rain. These areas are still worth protecting, but you may not want to spend a lot of money on a fence that may be damaged by floods.

Chapter 4 provides information on fencing, geotextile fabric, and other facilities supplies.

Managing your ponds and cattle together

A *farm pond* is a pool of water formed by a dam or pit. You can use it to supply drinking water for your cattle and to control erosion by capturing excess rain water that would otherwise wash away soil.

On hot summer days, cattle like to stand around in ponds trying to cool off. Doing so is unhealthy for your cattle and for your pond. The cattle will urinate and defecate in the same pond water that often serves as their drinking water. Standing around in the water stirs up sediment and makes the water muddy and may also cause the cattle's hooves to become soft and more susceptible to disease. And if mud accumulates around the pond, small calves can become trapped there.

For all the preceding reasons, most farm ponds should be completely fenced so cattle can't go around or in them. You can take advantage of gravity by using a drain pipe to bring water from the pond to a water tank at a lower elevation outside the fenced area. By completely fencing the area, you also keep cattle safe from falling through the ice in the winter.

If your pond is located in a flat area, however, gravity fed water may not be an option. For this situation, fence off the majority of the pond, allowing only one or two areas where the cattle can reach the pond to drink but are still limited in how far into the pond they can go. At the pond watering site, be sure to build a stable base with geotextile fabric and rock as you would for a stream crossing (see the earlier section "Building a stream crossing and drinking area").

Protecting riparian areas with flash grazing

Flash grazing refers to the process of grazing a paddock at a relatively high stocking density for a short period of time — typically two to four days — and then allowing the area to rest until the forage grows to 10 to 12 inches. Flash grazing is especially useful in riparian areas that are ordinarily fenced off from the cattle herd. By grazing these areas in quick, controlled situations, you can keep cattle from overgrazing the land around fragile stream beds and pond shores while still removing enough vegetation to maintain healthy, diverse forages.

For even more successful flash grazing, be sure to use these areas while they're fairly dry. Doing so keeps the cattle from getting too muddy and keeps their hooves from tearing up the plants and ground. Also avoid these areas in times of extreme heat because the cattle will congregate in the cool waters of the stream or pond. To draw the cattle away from the stream or pond, supply them with an alternative water source.

Chapter 7

Choosing and Buying Cattle

In This Chapter

- Picking out the best type of beef cattle for your farm
- Considering different sources of cattle
- Knowing what to consider before buying
- Transporting and caring for your new arrivals

If you're like us and enjoy raising livestock, it's hard to beat the anticipation and fun of searching for a brand-new herd or additions to an existing herd. Choosing new cattle is fun, but you need some bovine-buying knowledge so you purchase what's best for you and your farm.

This chapter takes a look at the main types of cattle most people raise: newborn calves, weanlings, show cattle, and bred cows. We first clue you in on the points to consider with each type. Next we discuss the various sources that sell cattle, explain what each has to offer, and point out the important things you should consider before buying. We end the chapter by offering suggestions for easing your new animals onto the farm.

Selecting the Right Cattle for You

During our years of raising cattle, we have bought or sold bovines of most every type. We have found that whether it's a bottle calf, weanling, show animal, or bred cow, each type has its benefits and drawbacks. To help you avoid having to experiment and discover all these pros and cons yourself, in this section, we discuss the important items to consider for each class. The animal you bring home should be a good match for your abilities and goals, so raising cattle will be a fun and potentially profitable experience.

Early start: Considering the fun and commitment of a bottle calf

A bottle calf may be a good choice for you if you want to raise cattle but don't have the funds for an expensive initial purchase. Bottle calves are most often sold *by the head,* meaning that the price is set for the animal regardless of its size.

Raising these young calves can also be a good project for the whole family to participate in. Youth programs such as 4-H and FFA (Future Farmers of America) may even have special classes at the fair where your kids can exhibit a bottle calf they helped care for. If you plan to sell the calf after starting it on solid feed, you won't need much living space for the animal. A compact shelter with solid sides and a roof to protect the calf from weather extremes and a small grassy or sandy lot should be sufficient. If you have the space, you can also raise the calf to market weight and sell it then or butcher it to stock your own freezer.

The small size of a calf may make it less intimidating to children, but even at 80 to 100 pounds, those little hooves can pack a wallop if you get kicked or stepped on!

Knowing where to find bottle calves

The potential sources for bottle calves include the following:

- **Dairy farms:** If you live near a dairy farm, you may have a good source of bottle calves right there. Because cows give birth all the time on these farms, they have a predictable, steady supply of calves, and the farmers are more likely to have a well-planned program for getting the newborns off to a good start. If you want to raise bottle calves on a large scale, you may even be able to work out an agreement with a dairy to grow out their female calves until they're ready to return to the milking herd.
- **Beef farms:** If you want to raise a beef bottle calf, ask cow/calf producers in your area to give you a call if they ever have a potential bottle calf. Just be prepared to take a calf on short notice! One common reason beef bottle calves become available is because a cow gives birth to twins. One out of every 100 to 250 births in beef cows produces twins. A cow can raise two calves, but it's physically demanding on her, and the calves usually don't thrive. When one of our mother cows births a set of twins and we're busy with the time demands of caring for the other new calves plus several other species of livestock, we're glad to sell one of the twins as a bottle calf to a good home.

  A heifer calf with a twin bull brother is often a freemartin. *Freemartins* have the XX chromosomes of females but, due to their exposure to male hormones while in the uterus, their reproductive systems don't develop properly. They're fine, healthy market animals, but they usually can't reproduce.

- **Feedlots:** Cattle in feedlots are typically housed in large pens while eating a high-energy diet to prepare them for butchering. Feedlot managers try to avoid pregnant heifers, but sometimes they have a surprise birth. These calves may be available at little to no cost, but keep in mind that because the mother wasn't eating a diet designed for a pregnant female, the calf may be small and weak.
- **Sale barns:** These facilities host public auctions of livestock, including cattle.

  Use caution when buying a bottle calf at a sale barn. Even when the calves are treated with plenty of care, sale barns are still stressful environments full of commotion and lots of potential disease-causing germs.

Caring for your calves

Yes, young calves are adorable and are an economical way of purchasing cattle, but they don't come without hard work and possible risk. Properly caring for a young calf means at least a twice-a-day commitment for 4 to 12 weeks. You have to bottle feed the calf milk replacer in the morning and evening until it consistently eats 1.5 to 2 pounds of solid feed a day.

Prepare your pocketbook and your emotions for potential sickness and even death along the journey of raising bottle calves. Compared to older, bigger animals, calves are more prone to health problems due to their immature immune systems and the stress of being born. These challenges are compounded when they're removed from or lose their mothers and their diets and environments are disrupted. Bottle calves may need treatment for *scours* (diarrhea), dehydration, or pneumonia. Sometimes all these problems can happen quickly and in rapid succession, so even with the best of care the calf may not survive.

To ensure a positive bottle calf rearing experience, keep these tips in mind:

- Make sure the calf consumes 2 quarts of colostrum in the first six hours after birth and another ½ gallon over the next six hours. (For more about colostrum, see Chapter 5.)
- If possible, continue having the calf nurse from its mother for the first three days.
- Have water and a grain mix or grain pellet especially formulated for calves available free-choice starting four days after birth.
- Select a calf with a good nursing instinct so you can feed it from a bottle with relative ease. A calf with a good nursing instinct will suck on your fingers when you put them in its mouth.
- Unless you're prepared to give extra TLC, avoid a calf with an unhealthy appearance. Unwell calves may have watery eyes, snotty noses, labored breathing, or wet, yellow manure.

- Check the navel of the calf. It should be clean with a dried-up umbilical cord. Flip to Chapter 12 for details on caring for the navel of a newborn calf.
- Keep the calf warm and dry during transport. A thick straw bed in the enclosed back of a truck is ideal.

Ready to grow: Exploring the benefits of a weaned calf

Even if you don't have the time or patience to raise a bottle calf (see the preceding section), you can still have the experience of growing out young cattle by starting with a weaned calf. A *weaned calf* is no longer nursing or drinking milk from a bottle; it has either started or is ready to start on solid feed. Calves raised naturally by a cow are weaned anywhere from 4 to 8 months of age. Bottle-reared calves are usually weaned 2 to 3 months after birth.

To avoid health problems and to get your animal off to a good start, it's best to buy *preconditioned calves.* No set standard exists of what it takes to be preconditioned, but these calves almost always have been away from their mothers for several weeks and are used to living on their own. They also have usually received at least one or two rounds of vaccinations. (We make recommendations for immunizations in Chapter 12.) Preconditioned males are more likely sold as steers instead of intact bulls, so you won't need to worry about the hassle of castration. The calves may be *bunk-broke,* meaning they're accustomed to eating from a feed bunk and not just nursing or grazing for nourishment.

Avoid calves that are freshly weaned. Being separated from their mothers and adjusting to a new diet is a stressful experience. If you add on the challenge of being transported and settling into a new home, these newly weaned calves are more likely to get sick than preconditioned calves.

Weaned calves are usually in plentiful supply about six to nine months following the main calving season. They're almost always sold by the pound.

Blue ribbons: Enjoying the rewards of a show calf

If you (or a family member) are interested in showing cattle, buying a steer or heifer that's especially bred and raised for exhibition may work for you. These animals should share many of the characteristics of the weaned calf: 6 to 9 months old, used to eating solid feed, recovered from the stress of weaning, immunized against common diseases, and castrated (if male). You can read more about weaned calves in the earlier section "Ready to grow: Exploring the benefits of a weaned calf."

The show calf has some additional traits and features beyond those of the run-of-the-mill weaned calf. Consider the following:

- They have superior body structure (we talk more about conformation later in the chapter).
- They're usually gentle and accustomed to being handled by people.
- They're often used to being tied up and trained to lead on a halter.
- They have been fed a special diet to maximize their physical potential and grow a shiny hair coat.
- Their parentage and genetics are often known. Based on their known genetics, you can then better predict their future performance.

Of course, all these show calf "extras" come with a price. Show calves are almost always sold by the head. Prices may start at levels a little over what's paid for generic, commodity, weaned calves and go up to thousands or even tens of thousands of dollars for animals that will be competitive at state or national shows.

A bun in the oven: Pondering the promise of a bred cow

Buying bred (pregnant) heifers or cows is like buying two for the price of one. Doing so can reap these benefits:

- **It's a quick way to increase the numbers of your calf crop.** These animals create gross income more quickly than raising your own replacement females. A bred cow produces a salable product in 12 to 24 months. If you raise your own replacement cow, it takes at least 4½ years from the time she is conceived to the time her first calf is ready to butcher.
- **It can hopefully bring superior genetics into your breeding program.** Buying a bred heifer or cow gives you the opportunity to make genetic upgrades to your herd. Not only will the calves produced by your new purchases be sired by a different bull than the rest of your calf crop, but the influence of the cow's bloodlines can impact your herd for years.

Be prudent when selecting bred females. Buy cows that were raised and thrived in an environment and management program that's similar to yours. A cow raised in cold, dry Montana on native pasture will have a difficult time adjusting to the high humidity and warm-season grasses found in the deep South and vice versa. Seeking out cattle close to home helps cut on trucking costs, too. When we look to buy new cows, we select females that were reared within 150 miles of our farm.

Why we like buying bred heifers

We have found the F1 cross females to be the best producers in our herd (for more on what it means to be an F1, see Chapter 3). But to raise our own F1 heifers, we would need to maintain two distinct purebred lines. We would also need a dedicated production plan to properly care for the unique needs of the growing heifer as she's prepared for pregnancy and during her early gestation. By buying bred heifers, we can instead focus on raising cattle strictly with the goal of beef production and on not diluting our efforts and resources trying to maintain a maternal line as well.

We have selected heifers over cows for a couple of reasons. In our part of the country (the Midwest), we have a bigger selection of bred heifers to choose from than cows in terms of breeds, quantity, and quality. We also feel like producers are less likely to part with their high-performing cows, and more often sell their lower producers. If they do sell a top-end cow, she will command a top-end price. Breeders often sell the majority of their bred heifer crop, so we get the opportunity (and the responsibility) of picking the best individuals. In some situations a breeder may sell his entire herd (maybe due to drought or personal circumstances); this would be a chance to buy bred females of all ages.

Most years we have ample pasture, so meeting the increased nutritional needs of the first calf heifers isn't a problem. We have them calve a month or two before the mature cows so we can give them extra attention. They also have more time to recover before rebreeding because we have them breed at the same time as the mature cows. So instead of needing to become pregnant within 90 days of giving birth, they have 120 to 150 days to conceive.

Even though a bred cow will have offspring to sell a year or so after you buy her, it will still take the sale of several calf crops to recoup her original purchase price. So make sure you choose cows that can be a productive part of your herd for many years.

You can buy bred cows and heifers throughout the year, but the selection is probably best in the four to five months proceeding the main calving season for your region. They're sold on a per animal basis, not by weight.

A *first-calf heifer* (a female raising a calf for the first time) needs extra care compared to a mature cow. Because the young female is still growing herself and caring for a calf at the same time, she needs higher-quality, more abundant feed than an older cow.

Deciding Where to Buy Your Cattle: A Bovine Shopper's Paradise

Where to begin! Considering that approximately 92 million head of cattle exist in the United States and that around 1 million producers involve themselves in at least one stage of beef production, you have a broad range of cattle to select from.

Multiple venues sell cattle, including cattle farms, consignment sales, test stations, sale barns, and the Internet. In this section, we run through the pros and cons of these different options so you know where to look for the best cattle at a fair price.

Going to the farm

We like buying cattle direct from the farm. Doing so gives us a chance to get to know the cattle breeder and determine whether his or her philosophies and production methods mesh with what we need to raise cattle successfully on our farm.

If you're just starting out raising cattle, visiting and buying from farms are good ways to find potential mentors. Even if you've been in the cattle business a long time, visiting a different farm can expose you to new or better ways of doing things.

When you buy cattle from a farm, you gain a lot of advantages:

- You can see whether the animals are growing and thriving on a simple pasture or whether they need fancy feed and expensive hay to look good.
- You can evaluate the true dispositions of the cattle. For instance, when you walk through the barns or pastures, you can determine whether the animals run for the hills or just look at you and then go about their business. Calm animals are easier to work with.
- You can probably see at least some of their relations. You can feel more confident about the way an animal will grow and perform if you can see what its mother or siblings are capable of doing.
- Buying directly from a farm reduces the potential for exposure to the germs and stress that occur when you buy cattle at off-farm locations, such as sale barns (more on these locations later in this chapter).

When you've chosen to buy from a farm, you can work with the seller to have healthcare or identification work done to the cattle before they come to your farm. When we buy feeder calves, for example, we have the seller vaccinate the calves according to our protocol so the animals have a chance to develop immunity before the stress of moving to a new home.

Of course, making a purchase directly from a farm also has its drawbacks. Consider the following:

- **Conflicts over terms and conditions can occur.** Unless you have a written contract detailing the terms and conditions of the sale, disappointments and conflicts over the performance of the animal may be difficult to resolve.
- **Setting a fair price can be more difficult.** If you don't like to negotiate, you may end up paying more than you should. When you buy cattle by the pound, you need to find certified, accurate scales for weighing.

  Take into account shrink when setting a per pound price. *Shrink* is the amount of weight animals lose when they don't have access to feed and water and due to the urine and manure lost when they're transported. Buyers at auctions figure that calves have at least 3 percent shrink. So these buyers can pay a bit more per pound because they aren't paying for the weight of stomach contents, urine, and manure. If you're buying cattle that don't have any shrink, you want to pay less than the standard auction price.
- **You risk becoming "barn blind."** Folks become "barn blind" when the best animal in the barn starts to look really, really good even when it isn't. If you're buying at the farm, it's critical to maintain perspective. Consider not only how the cattle look compared to their on-farm contemporaries, but also take into account how they stack up against your own animals or cattle from other farms.

Checking out consignment sales

Consignment sales aren't just for women's clothing and knickknacks. They can be fun, educational, and exciting for cattle buyers and sellers, too. They're most often sponsored by a breed association or state cattleman's organization. They may also be held in conjunction with a cattle show.

Here are some of the advantages of buying from a consignment sale:

- **A guarantee on performance:** Because they have the involvement of the independent third parties, consignment sales almost always offer a guarantee regarding the expected performance of the cattle. This guarantee can help give you more peace of mind knowing that an intermediary is available to help resolve any issues that arise between you and the seller. Also, the organizing sponsor may have health or physical performance requirements that all consignments must fulfill.
- **One-stop shopping:** Multiple cattle breeders enter animals in the sale, so you can check out the offerings of several farms all in one location. Consignment sale offerings can include all types of cattle: herd bulls, bred cows, cow/calf pairs, and show heifers or steers. Seeing many different farms' animals at once helps prevent barn blindness, because you have more animals to compare.

Some consignment sales are actually a combination of show and sales event. A judge evaluates and places the cattle from best to worst (at least in his or her opinion), and then the cattle go through the sale ring in that order.

- **A fair value:** Cattle at consignment sales usually sell for a realistic price. Because numerous sellers and bidders attend these sales, an animal has less of a chance of selling for way over or way under its actual value. However, you should always keep your wits about you so you don't bid beyond your budget.

Of course, some of the benefits that you get when buying off the farm (refer to the earlier section "Going to the farm" for details) aren't part of the consignment sale experience — and that can be a downfall. For instance, you won't get to see the environment the animal was raised in or have a chance to evaluate its relatives. The cattle sold at consignment sales also have to overcome the stress of traveling to the sale and fighting off any germ exposure there.

Taking advantage of test stations

A *test station* is a program, usually run by the state extension service, where bulls from various farms come to one location to *go on test.* The test is basically an assessment of the animal's performance over a 90- to 150-day period. They're all fed the same diet and measured for their growth, feed efficiency (how much feed is needed to gain a pound), reproductive soundness, and carcass merit (via ultrasound). After completing the test, the bulls are offered for sale through a public auction.

These test stations are a great way to compare bulls from different breeds and farms. You can get valuable performance information — such as carcass data or feed efficiency — that an individual breeder probably couldn't provide. Plus, to be eligible for the sale, bulls have to pass a fertility test and meet minimum performance requirements.

The feed rations at some bull test stations are designed to promote maximum growth. So if your feeding program differs greatly, the bull you purchase may not flourish in your environment. Bulls that excel on a high-energy, high-concentrate test diet may not perform as well on a diet higher in forages. Use care in transitioning the bull from his test station feedstuffs to your ration.

Visiting the local sale barn

Sale barns are the locations of public auctions where a multitude of sellers and buyers bring all species of livestock to sell to the highest bidder. They can be a viable source of cattle, but they can also be a source of a lot of bovine trouble. Luckily, as more marketing-savvy cattle producers have started working directly with the buyers instead of depending on the sale barn to act as a go-between, sale barns have had to offer more services and guarantees to stay in business. So local auctions often hold special sales where the participating sellers provide written confirmation verifying the background and health status of the cattle. Sometimes an on-site veterinarian will even give the sale cattle a specified set of vaccinations and treatments.

Some cattle producers aren't marketers. They want to raise cattle and let someone else take care of the selling part. Cattle from these types of operations are often sold at local sale barns. If you know where the cattle came from and feel comfortable about the way they were raised, these cattle can be good buys.

No matter what, always use caution when buying at a sale barn. Some unscrupulous sellers hide behind the cloak of anonymity provided by the sale barn auction to unload sick or poor-performing cattle. Even cattle that are well cared for may be exposed to stress and disease due to their sale barn experience. So be sure to quarantine cattle purchased at a sale barn and watch them closely for signs of illness.

Surfing the Internet

If you type "cattle for sale" into your Internet search engine, you'll get enough results to keep you reading for a year or more! You can find cattle of any number, shape, breed, or size listed for sale on the web.

Some farms have their own websites that describe their breeding programs and the animals they have for sale. Other sites serve as online classifieds, where you look for or post information about the cattle you want to purchase. Most breed associations devote part of their websites to selling animals; such sites serve as good places to start your search if you have a particular breed in mind.

Narrowing your search by using the Internet can help save you time and gas money, but it's still hard to beat actually seeing the live animal and meeting the person who raised it.

Making a Wise Purchase: What to Look For Before You Buy

Raising cattle is a big commitment of time and money. But it can be plenty of fun and financially rewarding, too. One of the keys to having a positive experience raising cattle is to make wise purchases when starting or expanding your herd. In the following sections, we review the important things you need to look at before you buy.

Noticing signs of a healthy versus unhealthy animal

We probably don't have to tell you, but when you're looking at a prospective bovine for purchase, you have to consider the animal's physical health. You want to give any animal you're considering buying a thorough looking over to gauge its health. Table 7-1 lists some things to look for and avoid.

Table 7-1 Signs of Healthy and Unhealthy Bovines

Physical Feature or Activity	*Healthy*	*Unhealthy*
General appearance	Alert with head up	Separated from the herd; slow to stand up
Eyes	Bright and clear with minimal discharge	Cloudy and watery with discharge present
Nose	Moist	Dry or with excessive mucus
Ears	Erect and moving	Droopy and stationery (except in droopy eared breeds)
Coat	Smooth and shiny; may be short or long depending on breed and season	Rough, dull, and unevenly grown
Breathing	Smooth and regular	Labored or shallow and rapid
Urine	Clear to light yellow in color	Dark yellow, brown, or red-tinged in color
Feces	Firm	Yellow, very dark brown or black, watery, or tar-like
Movement	Free and easy; weight on all four legs	Stiff or labored; won't distribute weight evenly
Appetite	Eats and drinks regularly	Reduced feed and water consumption
Cud chewing	6–8 hours daily	Minimal cud chewing

Avoiding genetic and bloodborne diseases

As you may imagine, you don't want to purchase an animal that's a known carrier of a genetic or contagious disease. If you do buy an animal that's a carrier, not only will you continue the spread of the disease, but you'll waste your money as well. A *genetic disease* is due to a defect in the DNA or genetic material of an animal and is passed from parents to offspring. A *contagious* disease is one that can be spread through the air, by direct contact, or via bodily excretions. We explain each in more detail in the following sections.

Examining genetic diseases

Because technology has become more advanced and less expensive, herdsmen can now identify several genetic diseases in their cattle with a simple blood or hair test. In fact, to be registered, many breeds of purebred cattle must undergo these tests or have parents with a known genetic profile. The results of these tests are presented on the animal's registration papers.

Crossbred cows rarely have these tests performed because, with their diverse genetic background, genetic diseases aren't usually a problem. However, popular crossbred bulls with semen available for purchase do have genetic disease tests performed. The results of the tests are listed in the semen catalog, or you can get the information from the owner of the bull stud that sells the semen.

Investigating contagious diseases

Depending on the region of the country where you live, you should test your cattle for a variety of contagious diseases. Some of the common diseases to check for are bovine viral diarrhea virus (BVDV), Johne's disease, brucellosis, tuberculosis, and bovine leukosis virus (BLV). Veterinarians can guide you in choosing what tests are right for your area. We discuss contagious diseases in more detail in Chapter 10.

Both the federal and state governments have rules regarding the required tests for certain contagious illnesses. Check with your local veterinarian to make sure you're complying with all regulations before transporting new cattle to your farm.

Assessing animal disposition

If you've ever had a beef animal with a bad disposition, you probably agree when we say a good temperament is one of the most important traits an animal can possess. A good temperament is important for many reasons, including the following:

- Working with calm cattle is better for your safety and the well-being of your cattle.
- Docile cattle are easier to work with and require simpler handling facilities.
- Cattle with a calm demeanor grow better, have increased fertility, and produce more tender meat.

Cattle with good dispositions should be mildly interested in you when you enter their barns or pastures, but they shouldn't flee to the far corners when you approach.

Avoid buying the following animals:

- Those that are constantly milling about their pens when you're nearby. These animals are showing signs of being nervous or wild.
- Those that are always watching you with their heads held high and ears perked.
- Those that don't stay with the herd.
- Those that show the slightest inclination to kick.

If having a quiet-natured individual is important to you, ask the seller to show how the animal reacts to being handled in close quarters. Also, watch how the seller acts around his or her cattle. Cattle that are handled calmly and consistently are more likely to act gently themselves. If the animal is being sold to you as a show animal, it should at least be accustomed to being tied up with a halter; ideally, it should be trained to quietly lead beside a person. Chapter 8 delves into handling cattle.

When buying a bull, have a frank discussion with the producer regarding the bull's disposition. Most bulls are rather slow and lazy (unless it's breeding season), but some of them can be quite aggressive or destructive.

Judging conformation

Assessing cattle *conformation* (their physical structure) is a skill that's part art and part science, and it takes time and practice to learn. The main things to look for when assessing conformation are muscling, size, and functional structure:

- **Muscling:** Although not every bovine is going to have a huge, rippling build, the animal should have some meat covering its bones. More muscle results in more salable pounds and more beef to eat.

- **Size:** Stay away from extremely small and extremely big animals. Little cattle don't produce as much weight to sell, and their small stature could be the sign of an unhealthy animal. Monstrously large cattle take more feed to maintain than their more moderately sized counterparts.
- **Functional structure:** Evaluate the skeleton and joints of your potential purchase. The animal needs to be able to move freely about so it can graze and feed itself. For breeding cattle, check that a bull has a large (greater than 33 centimeters), well-balanced scrotum, which is one indicator of fertility. Cows should have ample-sized udders with teats that are a moderate size for a calf to suckle.

A valuable way to improve your evaluation skills is to attend a cattle show and place the classes. Then listen to the official judge give his reasons for ranking the class in the order he did. You'll learn what traits are important to judges, how to describe what you're seeing, and how to compare and analyze the features of different cattle.

Understanding performance measurements and pedigrees

If the subjective business of evaluating cattle conformation leaves your objective side feeling deserted, you can use performance data to balance out. Performance measurements can be given as straightforward numbers, such as the birth weight of Calf #69 was 82 pounds.

However, most performance measurements are expressed as *expected progeny differences* (EPDs). EPDs provide estimates of the genetic value of an animal as a parent. Differences in the EPDs between two individuals predict the variation between their future offspring. EPDs are calculated by complex mathematical models and reported by breed associations.

Unless stated otherwise, EPDs can only be used to make comparisons within one breed. For example, comparing the EPDs of a Hereford bull to the EPDs of a Simmental bull is like comparing apples to oranges.

Some of the traits with EPD measurements include the following:

- **Calving ease:** This EPD measures the ease with which a sire's offspring are born. Using bulls with good calving ease should result in the need to assist fewer cows with difficult births.
- **Weights:** These measurements are taken for birth, weaning, and yearling weights. Calves with lower birth weights usually have an easier time being born. Cattle with higher weaning and yearling weights produce more pounds to sell.

- **Docility:** This EPD is a measurement of temperament.
- **Heifer pregnancy:** This measurement indicates the chance that a heifer will become pregnant in a specified amount of time.
- **Mature weight and height:** This EPD measures an animal's final size. Larger animals may need more feed than smaller animals.
- **Marbling:** This measurement determines the amount of fat within the muscle and can be indicative of the juiciness of the meat. This measure is determined with ultrasound for live animals and visually for carcasses.
- **Ribeye area:** This EPD is a measure in square inches of the size of the beef ribeye. It is determined with ultrasound for live animals and visually for carcasses.

Depending on the goals of your cattle-raising program, different EPDs and performance measurements may be important for you. If you're going to have cows give birth, you want to use a bull with good calving ease. If you're selling yearling cattle by the pound, large yearling weights will interest you. If you're raising cattle for beef, you should focus on the marbling and ribeye area traits.

EPDs and performance data are generally listed in the sale catalog for consignment sales and test station sales. Producers should be able to supply this information when you visit a farm as well. This data is also listed on an animal's registration paper or pedigree.

Evaluating customer service

Don't settle for doing business with just anyone. Ask for referrals from people you know and trust, such as your local extension agent or vet. (To find your local cooperative extension service, go to `www.csrees.usda.gov/Extension/USA-text.html`.) Look for producers who are active in local or state cattle associations. If someone takes the time to volunteer in beef promotion and education activities, they're more likely to spend time helping you and answering your questions.

Buy cattle from a producer who is transparent about the way she does things. Ask yourself these questions:

- When you visit the farm, is she willing and eager to show you all her cattle, or do you only get to see a select few?
- Are the cattle clearly and consistently identified?
- Is the producer open and honest about any health or genetic problems she has found in the herd?

- Does she have a breeding or performance guarantee for her sale cattle?
- Does the seller have a proven track record of satisfied customers?
- Are the bloodlines used in her herds similar to yours?
- Are the cattle well received in a variety of situations, such as bull tests, cattle shows, or consignment sales?

Arriving Home and Settling In

After choosing some new cattle to expand your herd, you face the task of hauling them to their new home and helping them adjust. You need to take care when transporting your cattle from the sale site to your farm. The less stressful the transition is for your new cattle, the more quickly they will adjust to their surroundings and begin to thrive.

Transporting your cattle

You have two options for hauling cattle: Do it yourself or hire someone else. Hauling your own cattle gives you greater control of the hauling process and doesn't limit you to someone else's schedule. However, owning a truck and trailer for a small number of cattle or for infrequent transportation is often not cost effective. Hiring an experienced hauler who has a good truck and trailer can be money well spent if he does the job right.

Regardless of who does the hauling, being moved to a new location can be stressful to cattle, so it's important to make the experience go as smoothly as possible. Here are some things to keep in mind:

- The truck and trailer should be in good repair, with sturdy construction, and it should be an adequate size to haul your load safely.
- The driver should have a spare tire for both the truck and trailer.
- The inside of the trailer should be clean and have a skid-resistant floor. To minimize the chance of disease transmission, a clean trailer is especially important if you're hiring someone else to haul your cattle.
- Trailer ventilation should be appropriate for the weather conditions. Common sense goes a long way toward keeping your cattle comfortable. Even in mild weather, they can become overheated in a poorly ventilated trailer.

Don't leave cattle loaded on a trailer that isn't moving during hot weather.

If you're dealing with extremely cold or wet conditions, ventilation is still important but should be at a level such that the cattle are protected from direct wind or overexposure during transport.

- Minimizing your cattle's travel time helps your cattle travel more comfortably.

Loading and unloading procedures are important for the successful transportation of your cattle. The loading and unloading pens need to be of sturdy construction and should be accessible (in all weather conditions) for large trucks and trailers to back up to. (Chapter 4 provides guidelines for designing a good handling facility.) Work with the cattle gently during loading and unloading and expect the same of anyone else involved (see Chapter 8 for proper cattle handling methods). Whooping and hollering isn't helpful and isn't good for the cattle. After the cattle are loaded, get them to their destination as soon as possible.

When your new cattle arrive at your farm, take time to inspect them. If you bought the cattle sight unseen, check to make sure the cattle are the size, class, and quality that you ordered. Additionally, look for any obvious health problems as discussed earlier in this chapter. Check for signs of injury, such as limping, swelling, or abrasions that may have occurred during the transport. Make notes of any problems you observe so you have a basis for treatments and for future evaluation when monitoring your cattle's recovery. If the cattle aren't as ordered, contact the seller and describe the deficiencies. If you can't reach a mutual resolution, consider refusing the load.

Adjusting cattle to their surroundings and feedstuffs

To get your cattle off to a healthy start, it's important to minimize stress when they arrive at your farm. To do this, you need secure facilities, a proper introductory diet, and easily accessible, clean, fresh water. The following are some tips for smoothing the transition:

- **Unload the cattle into a small, secure holding lot where they won't be tempted to run around recklessly looking for a place to escape.** Doing so allows them to quickly acclimate to their new surroundings, settle down, and move on to more important things like finding feed and water.
- **Have long-stemmed, weed-free, dust-free grass hay available, preferably in a bunk that allows all your new cattle room to eat together.** Place the feed bunk perpendicular to the fence. Cattle have a natural tendency to check the perimeter when put in a new enclosure. With the feed bunk perpendicular to the fence and blocking their path, they're more likely to stop and eat.

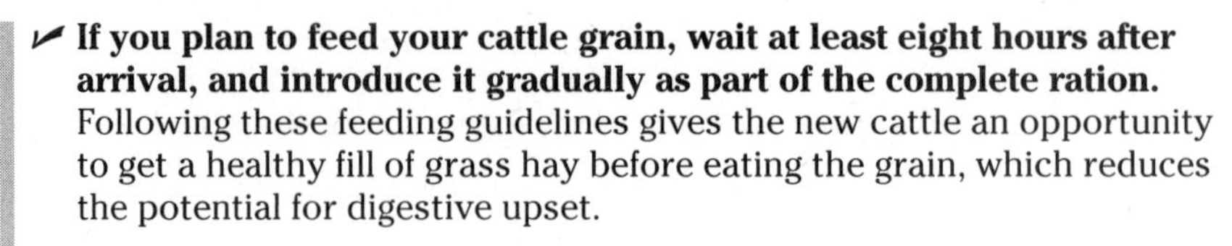

- **If you plan to feed your cattle grain, wait at least eight hours after arrival, and introduce it gradually as part of the complete ration.** Following these feeding guidelines gives the new cattle an opportunity to get a healthy fill of grass hay before eating the grain, which reduces the potential for digestive upset.

- **Don't overlook the importance of water.** Cattle that are hauled long distances or moved in warm temperatures can be very thirsty when they arrive at their destinations. Recently weaned cattle may be dehydrated due to stress before the trip. Also, some cattle are only accustomed to drinking from a stream or pond and aren't familiar with drinking from a tank. For all these reasons and because clean, fresh water is vitally important to your cattle's health, you need to ensure that your new arrivals can easily find the water source.

  Leave a hose running or set the float for continuous flow so the cattle can hear the running water. Also, as with the feed bunk, use the natural tendency of cattle to walk the fence of their new pen to your benefit. Place the cattle's water tank along the fence so they can easily find it during their initial inspection of the pen.

If you already have cattle or other livestock, quarantine new arrivals for 21 to 90 days. (The shorter time frame is enough for respiratory type diseases; if reproductive diseases are a potential concern, a longer quarantine is warranted.) *Quarantining* your new cattle involves keeping them separate from the rest of your animals. The quarantine period gives you an opportunity to observe your new cattle and improve your chances of identifying health problems that may exist without exposing the rest of your livestock. Diseases and other health problems can be passed through feed, water, traffic, and other means. The quarantine period also provides time for any vaccinations that were administered just prior to or soon after arrival of your new cattle to be efficacious.

Quarantine facilities should include a separate pen or pasture for your new cattle that doesn't allow fence line contact with the existing herd. Provide the new cattle with their own water source and feeding area as well. To the extent possible, don't allow vehicle or foot traffic to move between quarantined areas and the rest of the farm. If you can't eliminate traffic through quarantined areas, make sure you disinfect clothing, footwear, and vehicles after exposure to the quarantined cattle and their premises.

Part III

Cattle Handling, Health, and Breeding

In this part . . .

Here we get into the nitty-gritty of hands-on cattle care. Chapter 8 offers insight into cattle psychology and how to make handling your cattle a pleasure, not a pain. In Chapters 9 and 10, we discuss how to keep your animals healthy and explain what to do if sickness strikes despite your best efforts. We wrap up this part with Chapters 11 and 12, which provide a close look at how to best care for pregnant cows and their calves.

Chapter 8

Understanding and Properly Handling Cattle

In This Chapter

- Becoming familiar with bovine behavior
- Interpreting cattle body language and vocalizations
- Getting your cattle to work with, not against, you
- Handling bulls and new mothers

When raising cattle, you'll find that many things are out of your control, such as the drought that makes your pasture dry up, the judge who doesn't like your show calf's conformation, or the bad driver who takes out 20 feet of your new fence. However, you can control one important thing: how you handle your cattle.

Each stressful moment your cattle experience translates into reduced growth, increased susceptibility to disease, more chances of bad behavior in the future, and declining reproductive performance. If you have an understanding of how cattle think and react, you can work with them in ways that keep them calm and comfortable. In other words, by using proper handling techniques, you can get your cattle to cooperate with you instead of being forced to do things.

In this chapter, we cover the types of activities cattle like and don't like to do. We delve into improving bovine-to-human communication skills by helping you better understand cattle body language and vocalizations. We also give some specific pointers for moving and handling your cattle in a low-stress manner. Bulls and calving cows often require special handling, so the final section in this chapter gives you some ideas for working with these more temperamental animals.

Getting Acquainted with Bovine Behavior

Each species has its own unwritten set of rules about how each of the members needs to behave and interact with one another and their environment to survive and thrive. By understanding what makes cattle feel comfortable *and* uncomfortable and by being aware of some common activities, you can more easily work with your cattle and provide them with a more secure environment.

As you read this chapter and work with your own cattle, remember that all cattle are potentially dangerous. Even though cattle have been domesticated for hundreds of years, they're still large, powerful animals that may defend themselves or attack when they feel threatened. Remain alert at all times when working with cattle, and use extreme caution when children are around these animals.

Recognizing cattle preferences

Although each animal is unique, as a general rule, cattle personalities follow these patterns:

- When they're frightened, their first instinct is to flee, not fight.
- Abrupt noises and movements frighten them.
- They're curious animals that want to investigate new things or surroundings.
- They respond positively to kind treatment and food.
- They like to remain with the herd and don't like to be singled out.
- The herd is willing to move freely and will naturally string out, walking behind the leader.

Keep these behaviors in mind so you don't put your cattle in situations where their natural instincts are telling them to do the opposite of what you want them to do. For example, instead of pushing or forcing your cattle into a new barn or pasture, take advantage of their natural curiosity by leaving the gate open and letting the cattle move themselves. Also, if you need to work with a single animal, don't try to separate it from the herd out in a big pasture. It's usually easier and less stressful to corral or pen the entire herd or a large group and then remove the individual from the group.

Expecting cattle fighting

Cattle have a herd structure with a definite pecking order, starting with the boss bovine. Because cattle are prey animals, not predators, their natural instinct is to select a strong leader to help direct them when they feel threatened. The animals rank themselves all the way from the top of the dominance pile to the bottom. To establish this pecking order, the cattle sometimes fight.

Weaned calves and yearling cattle are the least likely to fight. However, in groups of mother cows or in groups of bulls, you may experience quite a few tussles until the herd structure gets established.

What does a bovine fight look like, you ask? It consists of a lot of pushing around with a few head butts thrown in. Cattle rarely kick at each other when fighting.

Because fighting is stressful to the cattle, try to minimize it as much as possible. Here are some ways to keep the peace:

- **Don't include horned cattle in your herd.** Polled (without horns) cattle rarely hurt each other when establishing dominance, but horned animals can do some damage.
- **Keep your herd's makeup consistent.** Unless absolutely necessary, don't mingle, split, and then remingle your herds. After a herd establishes its pecking order, fighting subsides greatly. Changing up the herd structure upsets the balance.
- **Avoid introducing single animals to the herd.** If you introduce just one animal at a time, they will get picked on and will be slow to assimilate.
- **Introduce the cattle in a controlled environment.** After the quarantine period is over, put the new cattle in a strong pen with open sides where the rest of the herd can see and touch them without coming in full physical contact. Doing so gives the animals a chance to get acquainted without getting too rowdy.
- **When new groups of females or calves come together, have them meet in a wide-open space with solid footing.** You don't want them getting pushed into something or slipping and falling.
- **House bulls individually and use strong fences and electrical wires to keep them apart.** Bulls can live together, but it works best when they're similar in size and age.

Dealing with riding or mounting cattle

When heifers or cows are ovulating, receptive to sexual activity, and likely to become pregnant, they're said to be in *standing heat* or *in heat* (estrus). This description derives from the way the females stand immobile while another cow, bull, or calf *rides,* or mounts, them. The cattle usually mount from behind, but calves also sometimes attempt to mount the head and neck instead. Of course, even though calves and females mount other cows, a pregnancy can occur only when a bull mounts and copulates with a female.

Riding is to be expected if you have females of breeding age, but it's tiring for the animal in heat and for the all the other animals riding her. Standing heat usually lasts for about 12 hours.

You can employ any of the following good animal husbandry practices to keep riding to a minimum:

- **Make sure your cows are in good physical condition.** A healthy, well-nourished cow is quicker to rebreed and won't keep coming in heat.
- **Make sure your bull is fertile and able to get the cow pregnant on the first try.** A bull won't ride a pregnant cow.
- **Castrate all male calves not destined for breeding stock.** *Steer calves* (male calves that have been castrated) are less interested in riding than bulls. After they become several months old, bull calves want to perpetually ride all the standing cows in the herd. This practice diverts energy that could be used for growth into useless riding. Also, the calves may suffer a physical injury by riding so frequently.

A female in heat is more easily upset than a cow that isn't in estrus. When a cow is around the time of her heat, be prepared for unexpected behavior.

Determining Whether Your Cattle Are Content

Even though cattle don't speak like humans, they can still most definitely communicate. Cattle transmit how they feel and what they want by how they position their bodies, move, or vocalize. You just need to slow down and pay attention to the signals the cattle are sending. In this section, we take a look at the two main ways cattle communicate: through body language and vocalization.

Deciphering body language: The head, the tail, and the stuff in between

Content cattle should be alert but not nervous. They should be with the herd, not off alone. Body language not only gives you insight into the mental state of your cattle but gives you information about their physical condition, too.

Talking heads

When cattle are calm, you typically won't notice anything unusual about their heads. They may show some mild interest in you and visually follow your movements across the barn or pasture. During these calm times, you may also see them chewing their cud. When a bovine is *chewing its cud* (a rhythmic moving of the jaws that grinds a small mat of partially digested and regurgitated plant material), the animal is feeling good and is comfortable in its surroundings. This behavior is a sign of a healthy and content animal. Ideally, you should see your cattle chewing their cud four to six hours a day.

If the positioning of the head and eyes conveys a state of being super-alert or watchful, use extra care around the animal. An elevated and erect head and neck are signs of a nervous cow. In addition, the animal's gaze may be fixed on you. A cow with a calf or an aggressive bull displaying this type of posture may charge (for more on these two special cases, see the later section "Safely Handling Breeding Stock"). Usually, other classes of cattle with this stance will bolt and attempt to run away, but you should never assume they won't come after you instead.

Cattle temperament is heritable and contagious. It takes only one bad cow or bull to be a negative influence on the entire herd. Don't tolerate cattle with bad attitudes. For your safety, sell these animals as culls (they will be used for meat) at the local sale barn.

Giving you an earful

Bovine ears can give you clues about the mood of the animal. The ears of a peaceful beef animal come straight out from the side of the head (unless of course the ears are naturally droopy like in the Brahman-influenced breeds; see Chapter 3). A moderate perking of the ears indicates curiosity within normal range. When the ears become fixed in position and tilted upward, you have a sign that the animal feels threatened and is compelled to be hypervigilant.

Ears that are swiveling and twitching a lot signal that the animal is irritated. The cause of the irritation may range from a biting fly to an uncomfortable halter or a funk (the animal may simply be in a bad mood and want to be left alone).

Knowing the tell-tail signs

The positioning of the tail can really telegraph an amazing amount of information about the attitude of your cattle. Consider the different tail positions:

- **Hanging straight down:** When they're grazing or walking, a tail hanging straight down means the cattle are relaxed and calm.
- **Tucked:** If the tail is tucked in between the hind legs, the animal may be cold, sick, or frightened. A young calf's tail is usually quite short, so if you have one with its tail pulled tightly into its body, take it as a sign that all is not well with the animal.
- **Flying like flags:** When cattle are in a playful mood or running for fun, the tail often is kinked or carried up over the back.
- **Standing straight out of the tailhead:** If the cattle are running and their tails are coming straight out of the tailhead, the cattle are probably agitated.

  An animal with its tail held stiffly out a few inches from the body isn't in a relaxed state. Depending on the situation, this positioning of the tail could mean the animal feels threatened or is investigating a new situation. It can also be a sign that the animal is preparing to mate or has recently mated.

Investigating other bodily cues

Even though the ends of the bovine body with the mobile head and tail are the two most expressive areas of the body, the middle region can tell you something about the well-being of your cattle, too. An animal that feels good and is getting enough to eat gives a big stretch and arching of the back when it stands up from the laying position.

Watching young calves stretch after rising from a rest period is especially important. One of the first signs that a calf may not be getting enough to eat is when it gets up without stretching and instead just stands with a hunched back. Here's what may be going on if you see this behavior:

- The calf may be sick and have a decreased appetite.
- It may not have learned how to nurse.
- The cow may have an infection in her udder (*mastitis*), which can stop milk flow or make the milk taste bad.

If you see a calf get up one time without stretching, make note to keep a close watch on it. Make the same calf get up again in an hour or two, and if it doesn't stretch at that time, check the cow's udder and the calf's ability to nurse. We provide more details about caring for the young calf in Chapters 11 and 12.

Translating the moo

As a general rule, cattle are quiet unless they have a problem. So you need to be able to differentiate the various sounds cattle make to know whether they need immediate help or can resolve an issue on their own. Describing cattle noises isn't the easiest task in the world, but we give it a shot in the following sections.

The Foghorn

Mother cows give the Foghorn, a loud and long call, when trying to rouse their calves. It goes from a low to a high pitch repeatedly with the same rhythm as a foghorn. Frequently mother cows bed down their calves to rest while they go out to graze. The calves are amazingly good about staying put until their mothers retrieve them and they hear this call.

Often a cow has to moo like this for several minutes before her calf responds. If this vocalization goes on for more than ten minutes and the cow is getting agitated, however, you may need to help the cow and calf find each other.

The Alarm

Calves use the Alarm, which is a loud, urgent distress vocalization, when they're hurt or scared. This type of moo puts all the cattle on alert, and all the mother cows run to the source of the noise. If you hear this call, you want to go hurrying to help, too.

The Alarm moo gets cows on edge and aggressive more than any other vocalization, so if you're the cause of the calf's distress (because you're giving it a vaccination, for example) be sure you're working with the calf in an area that's protected from the rest of the herd.

The Weaning Cry

The Weaning Cry moo is the crying sound a young calf makes when it's removed or weaned from its mother. This vocalization is often called *bawling;* it's persistent and repetitive, sort of like the irritating music you have to listen to while on hold. We talk more about how to practice low-stress weaning in Chapter 12, but even the best-laid weaning program is accompanied by these lonely moos from calves for several days. Even though the bawling can tug at your heartstrings and make your ears ring, it diminishes over just a few days.

The Ladies Man

Bulls give the Ladies Man call, a low-pitched, open-mouth call, to cows on the other side of a fence. The cows may be coming into estrus, or the bull may just want company.

When your bull starts making this vocalization, you want to be sure the fence is strong and the electric fence charger is on. Otherwise, you may face the unpleasant job of separating a moody bull from the cows.

Taking Advantage of Your Cattle's Natural Behaviors

If you want to be a good herdsman, you need to be aware of some innate behaviors that cattle have. If you know these behaviors and how cattle tend to respond to certain situations, you can control the setting to get the animals to act according to their own free will while still doing what you need them to. It's also important to realize what conditions cattle don't like so you can work to eliminate the conditions from their environment. We explain the most common behaviors that you need to be aware of in the following sections.

Managing the predator-prey relationship

Cattle are social animals and like to be in a group. This desire to be part of a larger crowd stems from the fact that cattle serve as the prey variable in the predator-prey equation. They're more comfortable as part of a group because being in a crowd offers some protection and help in watching for attacks.

So in your interactions with the cattle herd, you want to act in such a way that the cattle don't view you as a predator. Consider the following:

- **Keep noises to a minimum.** If the cattle can see you coming, approach silently because cattle don't like noises. However, if you're coming around a wall or up over a hill where the cattle can't see you, quietly alert them to your presence to distinguish yourself from the stealthy stalking of a predator. Our cattle are familiar with the sound of our voices because we talk to them quietly at feeding time, so we usually say something like "Easy calvies. We're coming up the hill."
- **Don't approach cattle from the rear.** A quick approach from behind is a predator-like action. By not approaching from the rear, you stay out of the cattle's *blind spot* (the area behind their tail end where they can't see). If your cattle are continually circling back to look at you instead of moving forward, you know you're in their blind spot.
- **Never approach cattle quickly while walking in a straight line.** Cattle vision is different from human vision. Because they have eyes on the sides of their head, they have almost 360-degree peripheral vision, which helps them spot predators coming from almost anywhere (except

their blind spot). The drawback with this eye placement is lack of depth perception. Cattle can't tell whether you're 15 feet away or twice that distance, so they don't know whether you're getting too close for comfort. To keep cattle calm, approach them at a slight zigzag angle rather than in a straight line.

Prey animals, like cattle, are good at concealing weakness or illness. They know that appearing unfit is like waving a red flag at a predator that says, "Come after me; I'm easy picking!" When your cattle no longer consider you a predator, they'll be more likely to show they're sick so you can better and more promptly care for them.

Using the flight zone to move cattle

The *flight* or *safety zone* is the space a bovine needs around it to feel safe and comfortable. It's a measure of how close you can get to an animal before it becomes uneasy and moves away. The size of the flight zone depends on the tameness of the animal. Cattle also have what's called the *point of balance,* which is the tipping point that determines whether the animal will move forward or backward. Figure 8-1 shows the flight zone and point of balance.

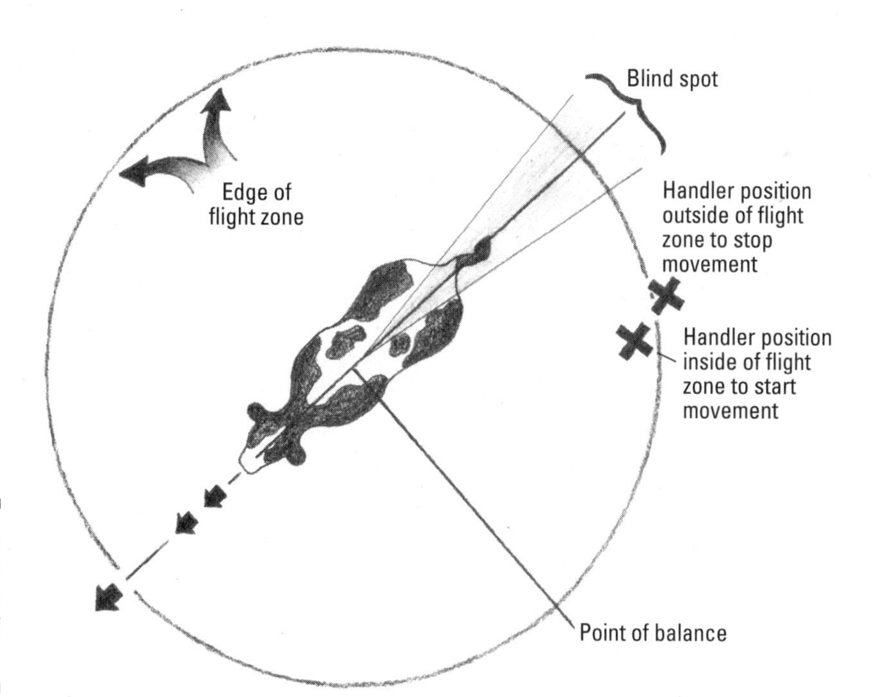

Figure 8-1: The cattle flight zone and point of balance.

Cattle that are worked with and handled on a daily basis (like show cattle) don't have a flight zone. It's often easiest to move them simply by leading them with a halter.

In the following sections, we show you how to use the flight zone and point of balance to move individual animals and herds. We also explain how to use gates to your advantage.

Moving individual animals

To successfully move an animal, you need to quietly enter its flight zone while staying out of its blind spot. (You can read about the blind spot in the earlier section "Managing the predator-prey relationship.") For cattle in a pen, the shoulder serves as the point of balance.

So if you're standing behind the point of balance and in the flight zone, the animal will move forward. If you're in front of the point of balance, the animal will stop or back up until you exit his flight zone.

By moving in and out of the flight zone and properly using the point of balance, you can move your cattle in a calm and controlled fashion.

Repositioning an entire herd

When working with a group of cattle, you need to think about the entire herd's flight zone and point of balance. The flight zone is no longer just the area one animal feels comfortable in; it's now the safety zone for the herd as a whole. For a herd of cattle out on pasture, the point of balance varies and may move as far forward as the eye of the animal walking at the front of the herd.

Here's a brief rundown of how to use the flight zone and point of balance to move a herd:

- **Move in and out of the flight zone to apply and remove pressure from the herd.** When the cattle are moving quietly in the right direction, ease up the pressure and move out of the flight zone. Figure 8-2 shows the placement and movement of two handlers in relation to the herd.
- **If the cattle slow down,** reenter the flight zone, approaching the herd at an angle and moving in the opposite direction of the flow of the cattle. They should speed up slightly to get by you.
- **If the cattle start to move too fast,** walk along beside the cattle at the front of the herd. Doing so slows them down as they position themselves at an angle instead of a straight line across from you. After they have decreased the pace, exit the flight zone as their reward for behaving appropriately.

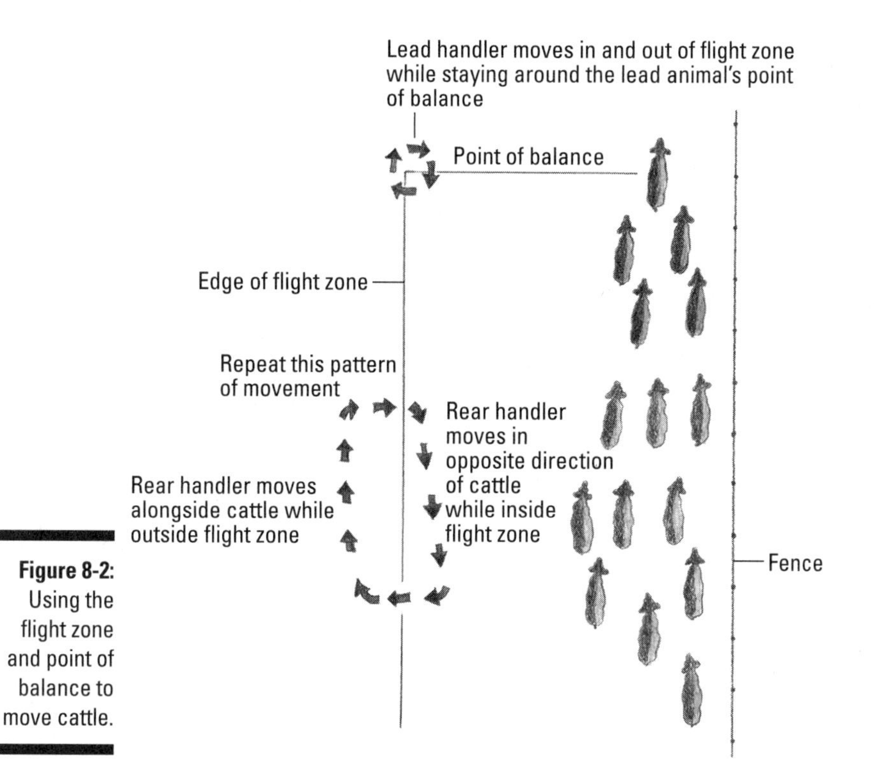

Figure 8-2: Using the flight zone and point of balance to move cattle.

Controlling movement through a gate

If you're moving cattle from pasture to pasture or need to sort some cattle in the handling facility, you can often control movement through a gate. Controlled movements help prevent the cattle from bunching up and getting hurt or bolting through the gate and becoming stressed out. Entering and leaving the flight zone helps manage the flow of cattle. Figure 8-3 shows where to position yourself and what movements to make.

Follow these tips to help your cattle make a smooth passage through a gate:

- **If you're at the gate opening, move in a parallel fashion to the cattle.** Moving in the opposite direction of the cattle causes them to walk by you, and moving alongside causes them to slow down.
- **Avoid the temptation to move perpendicular to the cattle.** For one thing, you'll be approaching them in a linear fashion, which is difficult for their limited depth perception to handle. Also, this type of movement could completely stop the flow of cattle and cause them to turn around.

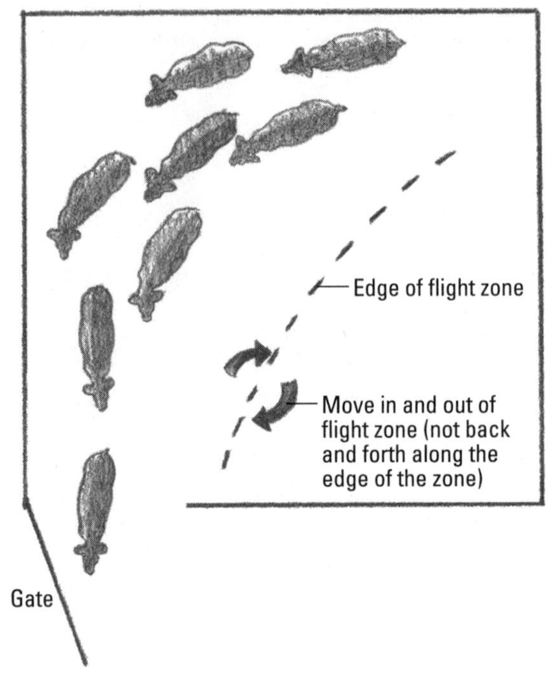

Figure 8-3: Handler movements and positions for getting cattle safely through a gate.

Working with cattle in a handling facility

A properly designed and maintained *handling facility* (a group of pens, alleyways, and equipment that provide a place for you to care for your livestock safely and in a low-stress manner) goes a long way toward making your cattle easy to handle. Chapter 4 goes into detail about the design and layout of a good cattle-handling setup, so here we focus on some tips and troubleshooting to help you best use your pens, chutes, and head gate.

Your cattle can give you feedback regarding their handling experience, so take the time to observe your animals. If they move at a steady pace through the pens and chutes with little to no vocalization, you're doing a good job of handling them. However, if they're upset, you may be guilty of one of the following:

- **Handling them before they've had time to adjust to their new surroundings:** Cattle need about 20 minutes to settle down after experiencing a stressful episode. Give them time to acclimate to the handling facility before you move them into the crowding pen or tub.
- **Crowding them:** If you get into an animal's flight zone, it will either express high stress levels or try to bolt over or around you. If you want the bovine to walk by you in an alley, stand to the side rather than behind it in the blind spot (which cramps an animal's style). Also, don't

fill pens more than half full. Cattle don't like to be packed together, and they often won't see the exit route because they're too stressed.

- **Working them too fast:** When you get in a rush, your cattle are more likely to sense your tension and get nervous. Give them time to find the opening to a chute or alley. Don't try to pressure them to move before they're ready.
- **Making too much noise:** The best handlers are silent when working cattle. Just by positioning their bodies properly in relation to the cattle, they can get the cattle to move as needed.

If your cattle aren't moving through the handling facility like they should after you have corrected your behavior, you may have problems with your equipment or facilities. Here are a few things to look for and fix:

- **High-pitched noises:** People can hear in the 1,000 to 3,000 Hz range, but cattle have the greatest sensitivity at 7,000 to 8,000 Hz.
- **Lights that cast reflections on puddles or metals:** Cattle get spooked by large contrasts in shades or color, so use absorbent ground covering to soak up the puddles.
- **Drafts:** Be sure no drafts are blowing into the faces of the cattle as they move through the handling facility, especially as they enter the chute. Drafts may cause the cattle to balk. Position your chute so it's sheltered from the wind.
- **Objects on the ground or fences:** Cattle don't like strange or unfamiliar things. They're even spooked by things as simple as a piece of garbage in the mud or a coat hanging on a fence.
- **Shadows on the walkways:** With their poor depth perception, the light and dark shadows cast by gates onto the ground frighten cattle. The black shadow stripes look like deep holes that cattle are afraid to walk over. Position lights so they're overhead and don't cause shadows to fall in the walkways.
- **Dark entrances or blinding sun:** Cattle move best through areas of equal lighting or from slightly darker to a bit brighter. They won't want to enter a dark barn or exit a head gate in the blazing sun. Use bright lights in your barn to help ease the transition from sunshine to inside or vice versa.

After they get going, cattle like to keep moving, but the action in the head gate and working chute is often done in stops and starts. To get the cattle walking again as one person releases the animal in the head gate, have another handler walk purposefully at a 45-degree angle toward the shoulder of the next animal in line. Seeing the bovine in front exit is a stimulus for the next one in line to move. And as the handler walks toward the animal, it wants to back up but can't because of the other cattle behind it. After the handler passes the point of balance, even the most hesitant cow can't help but pop forward into the holding chute.

Training your cattle

Cattle are definitely trainable. So it's up to you whether they learn good habits or bad habits. Here are some skills to teach your cattle:

- **Train them to have positive feelings about their initial experiences with people, events, and locations.** First impressions are everything — even with cattle! If they have positive experiences with new things, they'll fare much better. For example, to help give our new calves a good feeling about the corral and handling facility, we always bring them into that area for food treats. Two of the best food treats are grains like corn or high-quality alfalfa hay. When it comes time to work the cattle through the holding chute and head gate, they're comfortable and familiar with the space.
- **Teach them to come when called.** It's so much easier to have the cattle want to come to you instead of you having to go out and herd them up. It's also much less stressful for the cattle. You can train them to come to the beeping of a horn or the call of a voice.

Train the cattle when they're a bit hungry as opposed to times when the pastures are lush and abundant, because they'll be more interested in food treats when they don't have plenty of delectable forage. Stand just outside the cattle's flight zone when you call them so they can see the food treats you put out. After you've put the treat out, back up so the cattle feel comfortable approaching. They may need some practice to make the association between your signals and the treats, but after they figure it out, they won't forget.

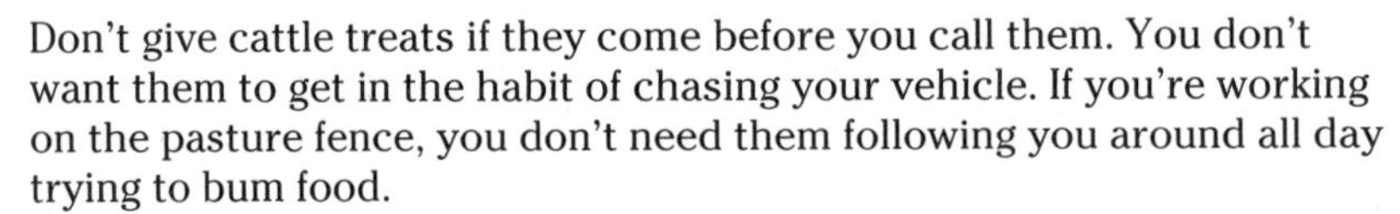
Don't give cattle treats if they come before you call them. You don't want them to get in the habit of chasing your vehicle. If you're working on the pasture fence, you don't need them following you around all day trying to bum food.

- **Teach them some eating-time manners.** This lesson is especially important for tame cattle you feed by hand in close quarters. For safety's sake, you need your cattle to stay back until you've dispensed the feed and moved out of the way. The best way to train them to wait for their supper is to take a long stick with you to the feed trough. If the cattle approach the trough before you're ready for them to, hit the ground (*not* the cattle) with the stick in front of them to keep them back. After you're ready, move back with the stick and give them a go-ahead verbal command.

Safely Handling Breeding Stock

The herd bull and the new mother and her calf can present special handling concerns. The bulls can be dangerous due to their massive size and basic instinct to be the protector of the herd. A cow with a young calf is hypersensitive because the momma knows her calf is prone to attack from predators. In this section, we give some tips for working with these two special classes of cattle.

Selecting the right equipment

When working with all types of cattle, but especially the potentially aggressive bull and new mother, you need to have the right equipment so you don't get hurt. Check out this list of must-haves:

- **Solid transportation:** If you need to bring the cattle in from the pasture, avoid going on foot. Instead, ride something solid to give you protection, such as a vehicle or an ATV.
- **Strong pens:** You can use these pens, which can be made of metal pipes, strong wooden boards, or solid sheets of metal, to safely contain the cattle when they're in a nervous or fighting mood.
- **Squeeze chute and head gate:** If you need to work with an adult animal, a squeeze chute and head gate are the safest way to control them (see Chapter 4 for more information about containment systems).

When working with a young calf, always put its mother in an adjacent but separate enclosure. That way the cow can still see her calf and won't go tearing up the barn and fences looking for it, but you have some protection from her. You can get a helper to hold the calf, or, if it's too boisterous, tie it up with a halter.

- **Nose rings:** Nose rings are helpful for bulls that you take to shows or for other situations where you lead them with halters. You have two different options for nose rings: a permanent ring that your vet can insert or a clip-on ring that's removable (it works like the old-fashioned clip-on earrings). The nose is sensitive, so a tug on the lead strap attached to the nose ring gets the attention of all but the most ornery bulls. The nose ring can also be snagged with a long, rigid piece of hooked wire to allow you to catch a tame bull and put a halter on him.

If you frequently depend on a nose ring to control your bull, you may have a sign that he's becoming too dangerous to handle. He still may be safe in a pasture setting but not for leading or showing at exhibitions.

Evaluating a bull's attitude and handling him

When working with bulls, keep in mind that each bull is different and that every bull can be dangerous. Before buying a bull, evaluate his temperament. You definitely don't want to buy (or be around) a bull that exhibits the threat display or actions of a direct threat.

A *threat display* is the stance a bull takes when he turns to give you a broadside view of his largest profile. He also puts his head down and shakes it from side to side, arches his neck, and his eyeballs may protrude. A bull may go into threat display mode when a human enters his flight zone (you can read about flight zones in the earlier section "Using the flight zone to move cattle"). Figure 8-4 shows the posture of a bull in the threat display mode.

If the bull feels that his display didn't cause the perceived threat to leave, he moves into the *direct threat* position. During a direct threat, the bull moves to face you head on. He lowers his head and hunches his shoulders. He may rub the ground with his head and paw the ground to send dirt flying. Move away from this type of situation immediately.

If you're cornered by a bull (or any bovine), avoid moving quickly; instead, back slowly away. Keep eye contact with the animal at all times. Turning and running is an invitation to be chased.

Of course, bulls sometimes exhibit some of the previous behaviors and aren't dangerous at all. Many bulls paw dirt up onto their backs to remove flies and bugs. They also toss their heads from side to side up over their shoulders to scatter flies. However, when you're working with 2,000 pounds of testosterone-infused muscle, it's better to be safe than sorry.

Figure 8-4: A bull exhibiting a threat display.

Working with mothers and baby calves

The best way to care for new mothers and their calves at birthing time is to do nothing! You may think that doing nothing seems counterintuitive, but, ideally, you will have done your work ahead of time. Your cows will be in good physical condition, so they're strong and fit, enabling them to deliver a calf without assistance. The calf will be sired by a calving ease bull (flip to Chapter 7 for details on calving ease bulls), so the delivery will be relatively simple and uncomplicated. Because the cow has had a good diet and the calf has gone through an easy labor, mom should produce plenty of milk, and the calf should be strong and ready to nurse.

Why is being hands-off during calving time so important? Labor is difficult for the cow and calf. Your "help" may actually be more of an interference, making the cow nervous and slowing the birthing process. In Chapter 11, we discuss how to determine whether your assistance is truly required. Also, after delivery, the cow and calf need time to bond. The new mother needs to lick the calf dry and imprint its smell as belonging to her.

Immediately after birth, the cow may be drawn more to the smell of the birthing fluids on the ground than to the calf itself. Therefore, if you need to move the calf from its birthing spot, be sure the cow follows the calf.

Chapter 9

Keeping Your Cattle Healthy

In This Chapter

- Monitoring the physical health of your cattle
- Vaccinating to prevent illness
- Scheduling annual health tasks
- Taking care of seasonal concerns
- Carrying out basic animal care procedures

For many farmers, one of the most rewarding parts of raising cattle is keeping them healthy and well cared for. On our farm, it gives us great pleasure to lean on the pasture fence and look out over a herd of fit and content cattle. Keeping a healthy herd has many benefits. For one, healthy cattle perform better and are more profitable for you. They also are easier to care for, have lower vet and medicine bills, grow faster, are more reproductively efficient, and produce higher-quality meat when they're in their prime.

In this chapter, we cover the many aspects of keeping your cattle healthy. We show you how to monitor the physical status of your animals and how to tend to them if they do become ill. We also prepare you to plan and perform routine preventive health tasks and procedures and provide a list of supplies to have on hand.

Observing Your Cattle to Recognize Illness Early

When aiming to keep your cattle well, an ounce of prevention is worth a pound of cure. Take the time every day to observe your cattle. Because you're much more likely to recognize an illness at the early, treatable stage if you know the difference in appearance between healthy and unhealthy cattle, we discuss some important indicators to look for when checking your cattle, including appetite, behavior, and excretory function.

Keeping an eye on appetite and behavior

One of the best indicators of illness is reduced feed consumption. For example, cattle with respiratory diseases often have decreased appetites two days before presenting with a high body temperature. Beyond appetite, cattle behavior also can provide insight on how well an animal is feeling. We discuss what to look for regarding feeding and behavior in the following sections.

Appetite

Feeding time is the best time to catch early changes in appetite. By taking a few minutes every day to observe your cattle during mealtimes, you can recognize their normal behaviors and know when they aren't feeling right.

In other words, for cattle you are limit feeding, don't just throw the feed in the trough and walk away. Instead, watch the animals for a bit. After you become familiar with your animals' normal feeding habits, you can later easily recognize which animals are exhibiting the following signs of poor appetite:

- Not coming to the feed bunk
- Slow to begin eating
- Coming to the trough but not eating

Assessing the appetites of grazing cattle or animals on a free-feed diet is more difficult. With these animals, look for signs of a full and rounded stomach. Cattle that haven't been drinking and eating normally appear gaunt and hollowed out in the abdomen.

Behavior

Paying attention to the attitude or behavior of your cattle, especially those you can't observe at the feed bunk, is a great way of judging their physical health. A normal, healthy animal is bright and alert and holds its head up to watch you. Depending on how tame it is, the animal may notice you and then return to grazing. Animals with a larger flight zone may move away from you. (Refer to Chapter 8 for more on flight zones.)

Two positive behaviors you want to watch for are the "stretch" and cud chewing. Cattle that feel good and that are getting plenty to eat give a big stretch when getting up from resting. They lower their heads, arch their backs, and may even curl their tails. Observing stretching in young calves is important because it's one of the easiest ways to tell whether they're well-nourished and healthy.

Similarly, the rhythmic moving of the jaws associated with cud chewing is a sign of a healthy and content animal. If cattle are stressed or feeling ill, they won't chew their cud. You should see your cattle chewing their cud for at least several hours a day. (Keep in mind, however, that it's normal for young calves that are mainly nursing to not chew their cud.)

Animals that are only slightly ill are hard to detect. Cattle in the early stages of illness or with a mild case often have slightly depressed attitudes. They may not be as alert or quick to react, but these differences will be subtle. As illness progresses, cattle often stand with their heads down and backs hunched up. Their ears droop and their stomachs become hollowed out. Wilder cattle may be slow to move away from you, and quiet cattle may not move at all.

As a disease advances, other signs of illness include

- Abnormal urine and feces (more on this in the next section)
- Coughing
- Difficulty breathing
- Discharge from eyes and nose
- Gaunt and shrunken appearance
- Isolation from the herd
- Little to no cud chewing
- Reluctance to stand
- Rough hair coat
- Slow movement

Checking urine and feces

Evaluating your cattle's urine and feces can provide information about their general health, excretory, and digestive function. A mature cow urinates and/or defecates 12–18 times a day, so you'll have plenty of opportunities for observations. On average, a single, mature beef animal produces around 80 pounds of feces and urine per day.

Try to look at urine as it leaves the body and before it soaks into the ground or bedding. Looking at feces soon after it hits the ground is best, but you can look at older piles, too, if you don't catch your animal in the act. Here are the general guidelines for checking urine and feces:

- **Urine:** Urine from healthy animals should be clear to light yellow in color. Your cattle should be able to pee without any discomfort or difficulty. In females, frequent production of small amounts of blood-tinged urine could be a sign of a urinary tract infection. For steers and some intact males, bloody urine may indicate urinary calculi.
- **Feces:** When assessing the manure of your weaned calves, yearlings, and mature animals, check out the following:

- **Color:** For grazing cattle, the manure is dark green in color. Cattle eating hay have brownish green feces. If cattle eat large amounts of grain, the manure is yellow-olive in color.
- **Consistency:** The consistency of feces depends on its water content, which is influenced by the moisture level of the feed. Normal manure varies a lot in its consistency but should be neither extremely fluid and watery nor dry and hard.
- **Content:** The content of the manure should be uniform and reflect all the different feedstuffs the animal has consumed. Expect to see some undigested feed but not a large amount.

The feces from calves are different from older animals. Chapter 12 has the scoop!

Get His Vitals, Stat: Measuring Physical Vital Signs

Vital signs are physiological measurements that allow you to assess your animal's ability to function. For cattle, the three key vital signs are temperature, pulse, and respiration rate. Vital signs provide a standardized way of checking your animal's health. If you have concerns about your bovine's well-being after a visual appraisal, taking its vital signs can help you further assess the health problem. We tell you everything you need to know, including when to take action, in the following sections.

Checking body temperature

Body temperature is the degree of heat of a living body, and it's a good indicator of health or illness. Table 9-1 details the normal and elevated temperatures in mature cattle.

Table 9-1 Body Temperatures of Mature Cattle

Physiological State	*Temperature Range (degrees F)*
Normal	100.4–103.1
Mild fever	103.2–104.6
Moderate fever	104.7–105.8
High fever	105.9–107.0
Very high fever	107.1–110.0

The normal temperature of calves is higher than that of adult cattle. A newborn calf's normal temperature is 101.4–104. As the calf ages, its temperature decreases but is still around 1 degree higher than mature cattle.

An animal's normal temperature can range by as much as 2 degrees. This fluctuation actually follows a predictable pattern. Cattle dissipate body heat mainly through respiration. Rather than wasting energy lowering their temperature during the day, their body temperatures rise during the day and go down overnight. Keep this pattern in mind when taking your cattle's temperature.

The lowest body temperature occurs in the early morning, making this an excellent time to check for a fever. At this time, you'll be less likely to have a false positive and reduce the chance of treating an animal for illness when its temperature is actually within the normal range.

Body temperatures rise outside of the normal pattern when cattle have an infection. The infection can be caused by a disease-causing organism, a wound, or trauma from giving birth. As the immune system works to fight off infection, the temperature rises. If the animal recovers on its own, the temperature goes back down. If the disease progresses, the temperature remains elevated and other signs of illness occur (refer to the earlier section "Keeping an eye on appetite and behavior" for details). In advanced cases of illness and as death approaches, the temperature falls below normal (*hypothermia*).

Of course, a bovine won't keep a thermometer in his mouth like you do, so you're going to have to take it rectally (fun, fun!). To make this task safe and easy, follow these steps:

1. **Check to make sure that your probe is securely attached to the digital readout or some string.**

   You don't want to lose the thermometer inside the rectum.

2. **Safely and securely restrain your animal.**

   See Chapters 4 and 8 for ideas on handling facilities and working with your cattle. Be careful to minimize stress and exercise before taking an animal's temperature because it will elevate the reading and make the animal feel even worse.

3. **Moisten the probe with lubricant prior to inserting it into the animal's rectum.**

   Vet and farm supply stores sell lubricants that are intended for animal use.

4. **Insert the probe and record the temperature after the digital readout stops fluctuating.**

   Follow the instructions that come with your thermometer to find out how far to insert the probe.

5. **Clean and store the thermometer as directed in the manufacturer's instructions.**

As you can imagine, most cattle don't like having their temperatures taken, so be sure to stay clear of any flying hooves.

Taking the pulse

The *pulse* is the push that blood makes against the wall of an artery with every heartbeat. It indicates the *heart rate,* the rate at which the heart is beating. Taking your animal's pulse gives you insight into the status of the circulatory system (heart, arteries, veins, capillaries, and blood).

The pulse can vary with age, time of day, exercise, and excitement. In adult cattle, the pulse can vary from 40 to 80 beats per minute. A newborn calf has a heart rate of 130 beats per minute. Just as when taking a temperature, be sure to take the pulse when the animal is calm.

When checking the pulse, you need to observe the following:

- **Frequency:** To determine the frequency of the pulse, count the number of thrusts against the artery per minute. A fever is often accompanied by a rapid pulse. Chronic diseases may slow the pulse.
- **Rhythm:** The proper rhythm of a pulse is steady beats at regular intervals. A variable rhythm may indicate an irregular heartbeat.
- **Quality**: Think of the quality of the pulse as the strength of the blood's push. A weak pulse may be a sign of a failing heart or reduced blood volume.

To feel the pulse in cattle, follow these steps:

1. **Place your first and second fingers on an artery close to the surface of the skin.**

   Choose a location where the artery is in soft tissue and where you can press it against a bone. The easiest artery to use is the external maxillary artery where it crosses over the mandible (or jaw bone). The artery is in front of the masseter muscle, which lies over the back of the jaw and moves the mouth open and shut. If you place your fingers flat on your animal's cheek in front of the masseter muscle and move them back and forth, you should detect the artery. It's a bit bigger than a drinking straw but much more pliable.

2. **Hold the artery steady underneath your two fingers and apply gentle pressure.**
3. **Count the pulses for a full minute to determine the rate, or frequency, correctly.**
4. **Hold your fingers on the artery for another 60 seconds to assess the rhythm and quality of the pulse.**

Observing respiration

Respiration is the process of breathing, in which oxygen is taken into the body (*inspiration*) and waste gases in the form of carbon dioxide leave the body (*expiration*).

The *respiratory system* consists of the nasal cavity, mouth, trachea, and lungs. Air enters the nose or mouth and travels through the trachea to the lungs. Many conditions negatively impact the respiratory system. It can be the primary site of a disease, or the respiratory system can fall victim to a problem that started elsewhere in the body. Because many illnesses can begin in or spread to the respiratory system, you need to know how to assess this system for any signs of trouble.

The *respiration rate* is the number of inspirations per minute. A mature cow at rest takes 10–30 breaths per minute. Calves have a rate of 30–55 breaths per minute. Unlike temperature and pulse, which require animal handling to measure, it's best to assess respiration rate from a distance. Because many factors like size, age, activity, nervousness, environmental temperature, pregnancy, and fullness of the stomach can impact respiration rate, it's most accurate to observe the animal in as natural of a state as possible.

When checking respiration, start at the nose and work your way back. Be sure to check the following body parts:

- **Nose:** It should be moist, not dry. Little to no discharge should be present. If you see discharge, make note of its color and consistency and whether any blood is present.
- **Mouth and trachea:** Look for abnormalities that impair breathing, such as cuts or sores in the mouth or abscesses blocking the trachea.
- **Chest and ribs:** Look for a regular rhythm as the animal is breathing. You should see no obvious straining. Instead, you want to see an easy expansion and relaxation of the ribs and abdominal wall.

Normal breathing should be silent. Listen at both the nose and chest area. Coughing, sneezing, wheezing, rattling, and moaning indicate a problem.

If you observe unexpected respiration rates, check a few other animals similar to your animal of interest. Doing so can provide you with a basis for comparison.

Deciding what to do for a sick animal

Cattle are quite hardy creatures. If they're living in a clean, low-stress environment with access to nutritious feedstuffs and water, they rarely get sick. If your animals do fall ill, however, be aware that you could be facing a complicated or serious problem.

So when should you take action (and what should you do) if you think your animal isn't feeling well? The answer to these important questions is . . . it depends! Factors influencing your decision should include the following:

- **The age of the animal:** Because they're more fragile, be much more aggressive in taking the vital signs of and initiating treatment in young calves (especially those less than a week old) than you would for a normally vigorous yearling. Follow these general guidelines:
    - **For calves:** If a calf doesn't perk up within six hours after you first notice a hint of illness, closely observe it for at least 20 minutes to further assess the problem. If you're still worried that the calf isn't feeling right, aggressively look for the cause of the problem. For details on caring for the young calf, see Chapter 12.
    - **For older cattle:** If the only indicator of a problem is that the animal is not eating normally, you can usually give it until the next feeding (12–24 hours) before taking action. If the animal is in obvious distress or showing signs of advanced illness (as described earlier in "Observing Your Cattle to Recognize Illness Early"), you should consult with your vet and begin treatment quickly.
- **The production stage of the animal:** For pregnant or lactating cows or bulls in breeding season, you need to be more assertive in diagnosing the problem than you would be for a nonbreeding bovine. The following list shows the reasons for each group:
    - **Pregnant cows:** Problems with the fetus, premature delivery, or digestive upset can happen quickly in the gestating female.
    - **Lactating cows:** The lactating cow can develop *mastitis* (infection of mammary tissue). Recovery from this condition has a much better prognosis if you get the cow started on antibiotic treatment right away. Chapter 11 gives more information on caring for your mother cow.
    - **Bulls in breeding season:** If your bull becomes ill or injured during breeding season, you need to get him feeling better fast. A sick or hurt bull may not feel like breeding cows or could become infertile.
- **External stressors the animal faces:** Your cattle can undergo stress inflicted by the following:
    - **Weather:** Extremes in weather, both hot and cold, can be hard on cattle. So can a swift and dramatic change in the temperature. Older cattle often graze right through rain showers, but a cold rain can bring about respiratory problems in a weak, young calf. If the rain doesn't cause problems for these small animals, the resulting mud might (calves can get stuck in heavy mud). Wind, snow, and ice are also hard on cattle of all ages and production levels.

  - **Change in diet:** Introducing new feeds to the diet or removing one can cause digestive upset or more serious problems like bloat.
  - **Transportation:** The stress of being transported makes animals more susceptible to disease. They may also be exposed to new germs if cattle outside your herd were hauled in the same trailer. Watch for injuries that may have occurred during travel as well.
  - **Introduction of new animals:** If you recently added new cattle to your herd, watch for two things: Keep a close eye on the new arrivals because they have gone through two of the stressful changes we just mentioned (transportation and likely change in diet). You also want to be mindful that even though your new animals were likely quarantined from the rest of your herd, potential disease-causing organisms may have still caused cross-contamination. (Chapter 7 provides steps for bringing home new animals safely.)

If your cattle experience any of these pressures and appear ill, be more rapid and aggressive with treatment than you would be if their living conditions were pleasant and consistent.

- **The number of animals affected:** If just one animal appears affected, you don't have to be quite as quick to act. You can take the time to observe the individual. When your whole herd is impacted, however, you must act much more quickly. You could be dealing with a severe, contagious, disease-causing organism. Or, it could be an issue with equipment, such as a broken water pipe that's causing an empty water tank.

If your animal is in obvious pain or distress, take action immediately.

When dealing with sick cattle, take the following steps:

1. **Make a written record of what you have observed about the sick cattle and any changes in the regular routine.**

   For instance, note any new feedstuffs, moves in pasture, or travels the cattle have experienced.

2. **Take the vital signs of the animal.**

   See "Get His Vitals, Stat: Measuring Physical Vital Signs" earlier in this chapter for details on how to do this.

3. **Isolate the sick animal from the rest of the herd to prevent the spread of disease.**

   Make sure the housing you choose is the best environment you can provide. If the weather is hot, run fans on the animal. During cold weather, keep drafts to a minimum. Provide plenty of clean, comfortable bedding.

4. **Provide clean, fresh water at all times. Also keep the cattle's diet consistent.**

Give the animal free access to its regular diet and provide high-quality grass/legume hay.

5. **Consult with your vet.**

   Use your vet's knowledge and expertise to identify the disease and what caused it. Your vet should advise you on the appropriate treatment or therapy.

6. **Continue to monitor the sick cattle to see whether additional treatment is needed.**

   Continue to record information and observations about the entire episode for future reference.

Vaccinating Your Animals

One of the best things you can do for the well-being of your cattle (and for your pocketbook) is to keep your animals properly vaccinated. *Vaccines* cause the body to build antibodies and develop immunity against certain diseases, just as if the animal had actually gotten sick and recovered from the disease. Cattle that aren't vaccinated or get vaccinated at the wrong time (jump to the next section, "Creating an Annual Healthcare Calendar" for ideas on correct timing of administration) are susceptible to costly and potentially deadly diseases. The main targets of vaccinations are respiratory and reproductive diseases. Vaccines come in two forms:

- **Killed vaccines:** This type of vaccine is made from killed disease-causing organisms, their parts, or byproducts. Compared to MLV products, killed vaccines have a slower onset of immunity and may not produce an immunity that's as long-lasting.
- **Modified live vaccines (MLVs):** This vaccine is made of a small quantity of viruses or bacteria that have been altered so they can't cause disease but can stimulate the animal to mount an immune response. MLV vaccines produce a wider range of protection than killed vaccines. However, they must be handled and mixed with more care.

Almost all vaccines are in a liquid form and given through injections. You can give the vaccines yourself or have your vet do it. See "Making injections a breeze" for the procedure for giving shots.

In most instances, you should be able to properly vaccinate an animal for less than $25 per year.

Creating an Annual Healthcare Calendar

Preventing disease is vital to profitable and enjoyable beef cattle production, so take the time to design and implement a herd health calendar (paper or electronic both work fine). To make sure you're providing your cattle with the proper care at the correct time, document on your calendar all the necessary healthcare tasks.

We can't provide you with a one-size-fits-all herd health calendar. After all, your cattle needs vary depending on the geographical region where you raise cattle, the type of cattle operation you have (cow/calf, feeder, or a bit of everything), and the kind of health problems you have encountered in the past.

So even though we cover the basic timeline for standard health and management tasks (by season) in the following sections, work with your vet and other consultants to design a program that fits the specific needs of your cattle. Chapters 10, 11, and 12 provide specifics about different diseases, preventions, and treatments.

Starting in pre-calving and calving season

This season includes the 30 to 45 days before calving as well as when the calves are being born. Here are the tasks you should have on your radar during this season:

- **Monitor the herd for nutritional needs.** Young, old, or thin cows need a diet that supplies enough nutrients for weight gain. Cows that are in good condition can receive a maintenance ration.
- **Clean and prepare *calving* (birthing) areas and supplies.**
- **Vaccinate pregnant cows and heifers 3–4 weeks before calving with enterotoxemia C & D toxoid and scours vaccines.**
- **Move expectant mothers to the birthing area about ten days before their due date.**
- **If possible, check your herd every 3–4 hours once calving season starts.** This check is especially important for *first-calf heifers* (females giving birth for the first time) because they're more likely to experience calving difficulties or not know how to care for a calf. Assist the calving process as necessary (see Chapter 11 for details).
- **Watch cows for post-birthing problems like uterine prolapse or retained placenta.**

- **Provide routine care to newborn calves.** Here are some of the main tasks (flip to Chapter 12 for more-detailed information):
  - Dip the navel of newborn calves in iodine to prevent infection. You can buy a nifty device that stores the iodine with an attached cup made for easy navel dipping.
  - Place ear tags in calves for identification purposes.
  - Make sure calves receive enough colostrum and are able to nurse.
  - Dehorn and castrate (if using bands for castration) calves a day or two after birth. Don't dehorn or castrate calves that are ill; wait for them to recover before doing any noncritical procedures.
  - Watch calves for any scours or respiratory problems.
- **Move cows and calves to clean, fresh pastures soon after calving.** Or move still-pregnant cows to a new birthing pasture, especially if the weather is mild, and let the pairs stay in the current pasture.
- **Perform breeding soundness exams on your bulls.** Also, vaccinate them for the following:
  - **Reproductive diseases:** Leptospirosis, vibriosis, and trichomoniasis (if recommended by your vet)
  - **Respiratory diseases:** IBR (infectious bovine rhinotracheitis), BVD (bovine virus diarrhea), PI3 (parainfluenza-3), and BRSV (bovine respiratory syncytial virus)
- **Monitor and vaccinate replacement heifers.** They should be at least 65 percent of their mature weight at the beginning of the breeding season. Vaccinate them for leptospirosis, vibriosis, IBR, and BVD.

Moving on to breeding season

For a cow to give birth annually (which is typical of productive females), she needs to become pregnant about 80 days after giving birth. Therefore, soon after calving season ends, breeding season starts. Perform these tasks during this season:

- **Start breeding yearling heifers 20 days before cows.** Doing so gives them extra time to rebreed the following year while staying in synch with the main cow herd.
- **Check eyes, teeth, feet, legs, and udders of cows for soundness.** Record this data. To keep up the performance of your herd, you may need to sell cows with physical problems in these areas. Good records help jog your memory when it comes time to make keep or sell decisions.

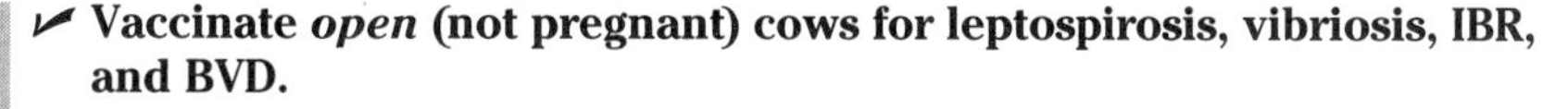

- **Vaccinate *open* (not pregnant) cows for leptospirosis, vibriosis, IBR, and BVD.**

  MLVs are only safe to use on open cows because they may cause abortions. If a female may be pregnant, use only killed vaccines.

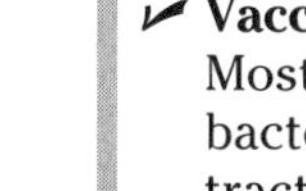

- **Vaccinate calves at 2–3 months of age against clostridial diseases.** Most of these vaccines confer immunity against seven different types of bacteria. These clostridial bacteria are found in the soil and digestive tracts of cattle. This vaccine is commonly called *7-way* or *7-way blackleg.* Consult your vet as to whether you should use a clostridial vaccine that also protects against red water, tetanus, or *H. somnus.* Also immunize calves against respiratory diseases, including IBR, BVD, PI3, and BRSV.

  Use only killed vaccines in calves that are nursing pregnant cows. The MLVs could move from the calf to its mother and cause an abortion.

- **Make sure cows are identified with clear and easy-to-read ear tags.**
- **Implant steer calves and heifers (that won't be used for breeding purposes) with growth promotants, if desired.** *Growth promotants* are small pellets placed in the ear that release hormones into the animal to increase lean muscle growth.
- **Vaccinate heifers that will be used as breeding stock for brucellosis.** The federal government mandates that this vaccination be done by an accredited vet during a specified age range (depending on the type of vaccine used).
- **If using artificial insemination (AI), make sure all handling facilities are ready and start heat detection.** After two rounds of AI, turn bulls in with cows.
- **Monitor bulls for libido, soundness, and condition.** Remove bulls from heifers after a 45- to 60-day breeding season. The mature cows should have a 60-day breeding season, including time spent for AI.

Entering into the preweaning and weaning season

Preweaning and weaning season is the time to prepare the calves for living without their mothers. It usually happens 5 to 8 months after they're born. During this time, you'll also complete tasks to help ensure another successful calving season for the following year. Here's what to focus on during this season:

- **Provide respiratory and clostridial booster vaccinations to calves 4–6 weeks before weaning.** You may also want to vaccinate for pasteurella at this time.

- **Dehorn all male calves and castrate those that won't be kept for breeding purposes if you haven't already done so.** Chapter 12 reviews different methods of castration and dehorning.
- **Weigh calves and record the information.** Make sure all calves have a form of identification.
- **Check all cows and heifers kept as breeding stock (*replacement heifers*) for pregnancy.**
- **Sell or develop a marketing plan for all open cows.**
- **Immunize cows against reproductive diseases if you didn't do so before the beginning of the breeding season.** Remember to use only killed vaccines in pregnant females.
- **Assess cow body condition so you can implement an appropriate feeding plan for your gestating cows.** Doing so helps ensure a healthy pregnancy and a vigorous calf. Check out Chapter 11 for more on body condition.
- **In some parts of the country or certain situations, you may also need to immunize for anthrax, anaplasmosis, and *E. coli* at this time.** Consult your vet for additional recommendations, too.

Preventing Seasonal Diseases and Issues

Some animal care issues correlate with the different seasons as opposed to the cattle production cycle. They include internal and external parasite control, pinkeye control, specific vitamin and mineral needs, and protection from weather extremes. To keep your cattle comfortable and healthy, incorporate the surveillance and treatment of these concerns into your overall animal healthcare program. Chapter 10 provides details on treatment options for the diseases in this section.

Being a bad host to internal and external parasites

Parasites are living creatures that require a host (such as your cattle) to live in or on at some point in their life cycles. Cattle infected with parasites have reduced growth and reproductive performance, making them less profitable. They also suffer from decreased feed efficiency, which leads to higher feed bills. To top it all off, parasites cause cattle discomfort and stress. The following sections provide a rundown of both internal and external parasites.

Internal parasites

Internal parasites include stomach worms, lungworms, liver flukes, coccidian, intestinal worms, and tape worms. They all spend part of their lives, as larva and egg-laying adults, inside a cow. The eggs are shed in manure where they hatch into larvae. The larvae crawl up the blades of grass in the pasture, and then when cattle graze they ingest the larvae and the cycle continues.

No single formula for controlling internal parasites exists. However, keep the following in mind:

- Mature cows have more resistance to worms than young animals.
- Animals in dirt lots (without access to grass) have a lower parasite load than pastured cattle.
- You can assume if your cattle have grazed they'll have some degree of infestation.
- Parasite contamination is highest in heavily grazed, short pastures.
- Parasites can live through freezing temperatures either by hibernating in the host or burrowing into the soil.

Partner with your vet to put together the deworming program that works best for your herd. Here's an initial plan that you can tweak to match your needs:

- **Treat mother cows in the spring.** If they haven't calved yet, be sure to choose a product safe for use in pregnant females. Treatment at this time is a double hit. For one thing, the levels of parasites are highest in this season, so you're attacking the greatest numbers of worms. Also, the hormonal changes associated with late pregnancy and calving seem to make the dormant worms in cows become more active and susceptible to treatment.
- **Deworm bulls in the spring and fall.** Bulls don't have as much resistance to internal parasites as cows, so they benefit from the second treatment.
- **Administer anthelmintics (dewormers) to calves at 3–4 months of age and again at weaning.** If your cattle are going off pasture into a feedlot after this time, no more treatments are necessary. Treat animals that will continue living on pasture in the spring and fall until mature.
- **Treat any purchased cattle, regardless of age, that you bring to your farm for worms unless they were recently treated and off pasture until arriving at your farm.** You want to bring only new cattle to your farm, not new worms!

External parasites

As their name implies, *external parasites* are found (for the most part) on the outside of your animal. The variety of external parasites — flies, ticks, grubs, lice, and mites — live at different times, so your cattle could be under assault all year long. Consider the following external parasites:

- **Flies:** The number of flies present on the side of an animal is a good indicator of when to start fly control. If you see 50 flies or more on each of your animals, take action. It only takes a few weeks for the fly population to double and decrease cattle performance. Flies are annoying to your cattle and can cause pinkeye (see the later section "Controlling pinkeye"). Depending on your region, flies can be a problem starting in May or June and can continue through September or October. In mild climates, flies can be a concern the entire year.

  If you use ear tags with insecticide for fly control, remove the tags at the end of fly season. Doing so helps slow the fly's ability to develop resistance to the insecticides.

- **Ticks:** You can see these small, blood-sucking parasites attached to your cattle around the tailhead or in the ear. Ticks also burrow into loose folds of skin. Cattle with heavy infestations may lose weight and become anemic. The ticks' bites are irritating to the animals, compelling them to rub and scratch. Be prepared to treat for ticks during times of moisture, warmth, and high humidity.
- **Heel flies:** These parasites lay their eggs on the hair of cattle's heels. The larvae that hatch from the heel flies' eggs are *grubs.* The grubs migrate through the skin into the animal and overwinter near the throat or backbone. You can actually see the *warbles,* or cysts, the grub larvae make along the throat and spine. (Chapter 17 discusses how you can determine whether your cattle are bothered by heel flies.) Confer with your vet for the optimum date for treating grubs. You want to target the time after the heel flies are no longer active but before the grubs have migrated to the throat or back. Depending on your location, properly timed application could be August, September, or October.

  Don't treat for grubs outside of your vet's recommended time frame. Paralysis or choking could result due to the grubs' location near the spinal cord and esophagus.

- **Lice:** These parasites are an annoying problem for cattle in the winter. They can even cause weight loss in severe cases. Cattle that have a louse infection present with hair loss around the neck and tailhead and with severe itching. If you part the hair at the infested area, you can see the lice moving through the hair or with their mouths embedded in the skin.

To completely eliminate the lice, you need to treat with a product that has a duration of action of at least two weeks or give two treatments 14 to 18 days apart. This time frame allows the treatments to kill both the adults and larvae.

- **Mites:** *Mites* are tiny relatives of spiders; they cause the infection called *mange.* This infection spreads easily and makes cattle quite uncomfortable. Cattle with mites scratch a lot and have areas of thickened skin. To differentiate between lice and mites, your vet can examine a skin scraping under the microscope. Mites are mainly problematic during late fall, winter, and early spring.

Controlling pinkeye

Pinkeye is a contagious bacterial infection of the eye that's characterized in the early stage by watering of the eye and squinting. It is painful, decreases the performance of the afflicted cattle, and can lead to blindness.

Pinkeye can occur any time of the year, but it's most frequent in the late spring, summer, and early fall when face flies are more common. The *face fly* is an insect that feeds on the secretions of the eye and, in doing so, irritates the surface of the eye, making it more prone to infection. The fly can also transport the disease-causing bacteria on its body, thereby quickly spreading the infection.

Start antibiotic treatment at the first indication of disease. If your herd has had multiple cases of pinkeye in the past, vaccinate your calves at least 30 days before the start of fly season.

Beefing up seasonal nutritional needs

As you complete your month-by-month health calendar (which we discuss in the earlier section "Creating an Annual Healthcare Calendar"), include a reminder to feed the following:

- **Vitamin A:** Vitamin A is particularly important during those times of the year when you're feeding poor-quality hay and supplementing with corn, and no green grass is available. During this season, your cattle need the extra vitamin A to maintain healthy skin, body linings, and night vision.
- **Magnesium:** Several weeks before cattle start grazing lush spring grasses, be sure to start feeding them a multi-mineral supplement with high levels of magnesium to prevent the potentially fatal metabolic disease called *grass tetany.* (See Chapter 5 for specific recommendations.)

Keeping cattle comfortable during extreme weather

Cattle need extra care during seasons of harsh weather to ensure that they stay well and that their performance doesn't suffer. Environmental challenges can come in the form of cold, wet, and windy weather or extremely hot weather. We describe how to deal with both in the following sections.

Coping with old man winter

When dealing with wintery weather, keep in mind the lower *critical temperature point.* This point is reached at the lower end of the animal's temperature comfort zone. After the temperature falls below this comfort zone, the animal has to increase its rate of heat production, which requires more energy. As a result, you must feed it more. Performance also decreases as the ambient temperature goes below the critical temperature point.

The critical temperature point for an animal depends on the following factors:

- **Coat:** If an animal has a thick winter coat of hair, its critical temperature point is in the 20- to 30-degree Fahrenheit range. If you have animals with thin coats, you need to increase their amount of feed 1 percent for every 1-degree drop in temperature. Cattle with thick coats only need a 0.5 percent feed increase for every 1-degree decrease.
- **Moisture conditions:** In dry conditions, the hair coat stays fluffy and traps body heat. However, after it becomes wet due to rain or snow, it loses its insulating quality and the animal's critical temperature hovers around 59 degrees.
- **Fatness, or body condition:** Fat is a good insulator, so animals with a good body condition can withstand colder temperatures in the 15- to 25-degree Fahrenheit range. In extremely cold weather, fat is metabolized to provide energy. See Chapter 11 to find out how to assess body condition.
- **Age and size:** Mature animals are better able to deal with cold weather than calves because they usually have more insulating fat than calves. Calves also have a high surface area to weight ratio, so their body heat dissipates more quickly.
- **Duration of cold weather:** Cattle can handle brief periods of inclement weather easier than they can deal with long periods of the same bad weather.
- **Amount of wind:** The faster the wind, the colder the body feels. You must take into account the wind chill factor when figuring the difference between the animal's critical temperature point and the temperature

actually felt by its body. For instance, if the air temperature is 30 degrees, but the wind chill makes it feel like 0 degrees, your cattle need about 30 percent more feed than they would if no wind were present. Windbreaks are an effective way to offset the impact of wind. (See Chapter 4 for more info on creating wind breaks.)

Contending with hot and humid conditions

Hot and humid weather can decrease reproductive and growth performance, so be sure to check on your cattle regularly during these difficult conditions. Some steps you can take to help cattle deal with the heat and humidity include the following:

- **Provide shade in the form of trees, buildings, or sunshades.** Chapter 4 gives you details about how to effectively use these shade options.
- **Improve airflow in barns by opening windows for cross-ventilation and by using livestock fans.** A nice breeze makes the animals more comfortable.
- **Provide plenty of cool, fresh water.** Water intake can nearly double for cattle during hot, humid weather. If your cattle are tame you can spray them directly with cool water. For more about cattle water requirements, head to Chapter 5.

Mastering Basic Animal Husbandry Skills

To care for your cattle correctly, promptly, and economically, be ready to master some basic animal husbandry skills. After all, for some routine care, straightforward problems, or urgent matters, it just doesn't make sense to hire a vet (or to wait for one!). By being able to give shots, put in tags, and care for wounds, you can address many of the health needs of your cattle.

In this section, we give you a quick rundown on how to do all these tasks. But, first, we get you prepared for your adventures in animal husbandry by showing you how to create a first-aid kit to have on hand at all times.

Injured cattle may be frightened or stressed. Don't expect them to behave like they usually do. Handle them with caution and care.

First things first: Stocking your first-aid kit

When you're raising livestock, you're bound to have unexpected emergencies sooner or later. To be prepared and able to care for your cattle quickly and effectively, make sure you have the following tools in your first-aid kit:

- **Halter and rope:** Before you can tend to your injured animal, you need to be sure it's safely restrained. A halter and rope are often useful for this task. Refer to Chapters 4 and 8 for ways to control your animal.
- **Strong, dependable flashlight:** A flashlight is important because emergencies can happen day or night. You may even consider having an extra set of batteries just in case.
- **Disposable latex gloves:** Gloves protect you and your animal from spreading infection or disease.
- **Wire cutters:** This tool helps you remove wire or fencing that may have wrapped around an animal's foot or leg.
- **Needle-nosed pliers:** Use this tool to pull out embedded objects like nails, wire, or large thorns.
- **Gauze pads:** When bandaging wounds, press gauze pads against bleeding injuries and use them to clean wounds. Both small pads, such as the 4-x-4-inch size, and larger ones are helpful.
- **Tape:** You can use tape for many purposes, including holding bandages, wrapping sprains, and holding splints into place. Stretchy vet wrap, 1-inch medical tape, and duct tape all come in handy.
- **Sharp, strong scissors:** Having scissors on hand is important for when you need to cut tape and other items.
- **Rubbing alcohol:** Use rubbing alcohol to disinfect healthcare equipment or the animal's skin before performing a procedure, such as a blood draw or ear tag insertion.
- **Sterile saline:** Saline is useful for flushing material from the eye or from wounds.
- **Water-soluble lubricant:** You can use lubricant to aid in inserting an esophageal tube or thermometer.
- **Wound spray:** You can use this spray to disinfect cuts and scrapes. It also protects the wound and repels flies. Just be sure it's approved for use on food animals.
- **Syringes:** You need syringes for giving shots (2, 5, 10, and 20cc) and for flushing tissues or eyes (30–60cc).
- **Needles:** Stock your kit with needles for giving injections. The 16-gauge size is most common.
- **Thermometer:** Yup. Cattle need their temperatures taken, too.
- **Antibiotic ointment:** Have formulations for both the eye and skin on hand at all times. Just be sure the ones you choose are safe for use on food animals.
- **Epinephrine:** This medication is used to treat anaphylactic shock. It's often used on animals that have severe allergic reactions to vaccines.

Pack your first-aid kit in a sturdy, waterproof container, such as a plastic tote or tackle box. Inside the box, tape a list of important contact names and numbers for people who can help you in an emergency, such as your vet, capable neighbors, and so on. Make sure that everyone helping to care for your cattle knows the location of the first-aid kit. Also put together a compact kit to carry in your farm truck and one to take to cattle shows.

Making injections a breeze

In order to properly administer vaccines and other medications to your cattle, you need to know how to give an injection. Before attempting to give your cattle a shot, make sure you have them safely restrained in a chute with a head gate (see Chapter 4 for information about cattle restraint).

Every medication comes with a package insert that provides dosing and administration directions as well as any safety warnings and withdrawal times. *Withdrawal time* refers to the amount of time that must pass between an animal getting a shot and the animal being butchered for human food consumption.

Read the package insert to ensure that you give your cattle the right amount of medicine, that you administer it properly, and that you follow the warnings and withdrawal times.

The most common routes of administration for injections are

- **Subcutaneous:** Under the skin
- **Intramuscular**: Into the muscle
- **Intravenous**: Into the vein

We describe each of these and how to administer them in the following sections.

Subcutaneous shots

When giving a subcutaneous, or sub Q, shot, you're aiming to go under the skin but not into the muscle. So you want to use a needle that's ¾-inch in length (enough to get through the hide but not much farther) and 16-gauge. (*Gauge* refers to the diameter of the needle. The larger the gauge number, the smaller the needle.)

To give a sub Q shot, follow these steps:

1. **Grasp the skin on the side of the animal's neck in the region shown in Figure 9-1.**
2. **Pull the loose hide enough to slightly "tent" the skin, as shown in Figure 9-2.**

3. **Insert the needle into the tented area, and then inject the medication in the loose tissue beneath the skin.**

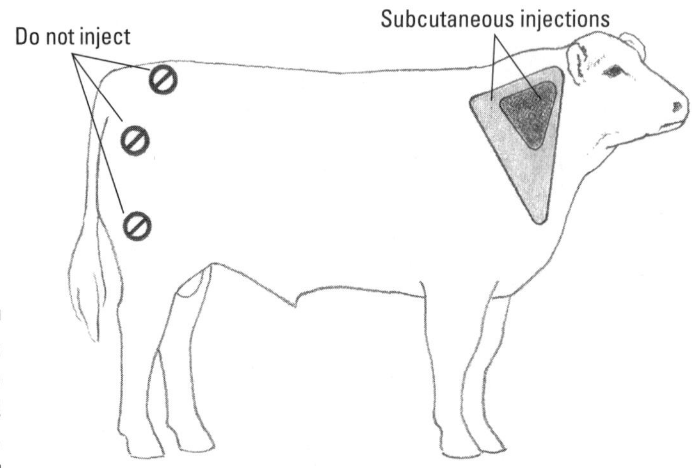

Figure 9-1: Proper site for a sub Q injection.

Figure 9-2: Tent the skin and insert the needle for a sub Q injection.

If the animal isn't restrained sufficiently to allow you to safely grasp and tent the skin while giving the shot, you can generally get good sub Q placement using a shorter needle (½-inch) and inserting it at an angle that prevents material being injected into the muscle.

Intramuscular shots

When giving an intramuscular, or IM, shot, you want the medication to go directly into the muscle tissue. So for this injection, use a 1- to 1½-inch, 16-gauge needle. You make the injection in the heavily muscled region of the neck.

Because IM injections cause tissue damage, don't give one in any area other than the neck, and don't give more than 10cc of product per IM injection site. In fact, avoid using IM medications if other options are available. If more than 10cc of product is needed, the remaining portion should be given by a second IM injection on the animal's other side.

Intravenous shots

Intravenous, or IV, dosing can initially be challenging. However, with a little direction and practice, it becomes much easier. We encourage you to get instruction from a veterinarian before attempting an IV injection.

Keeping records of your injections

After giving injections like vaccines or antibiotics, recording information regarding the shot is important. Good records help you stay on top of administering the right medications to the right animals at the right times. Record the following information:

- Name of injection and route of delivery (IM, sub Q, or IV)
- Manufacturer and supplier of injection
- Serial number of injection (in case of recall or product failure)
- Amount given and date given
- Identification of cattle receiving injection
- Animal's reaction to the injection, such as whether the illness was cured or the animal had a bad reaction to a vaccine

If your cattle respond appropriately to the injection, you can look at these records and repeat the same procedure for future health work. If your cattle continue to have problems after treatment or vaccination, your records serve as a good starting point for figuring out what's going wrong.

Hi! My name is Daisy: Attaching identifying ear tags

Ear tags are an easy way to identify your animal, and we recommend that you use them. Even if you recognize all your cattle without looking at the tags, a part-time helper may not! Here are the proper steps to tagging your cattle:

1. **Write the desired information on the tag with a specialized tag marker.**

   Allow the ink time to dry before tagging. You can match calf and mother ear tag numbers or write info about the calf's sire or date of birth.

2. **Secure your animal in a head gate to prevent injury to human or beast.**

   Besides keeping everyone safe, securing the animal also aids in correct tag placement.

3. **Using rubbing alcohol or a disinfectant solution like chlorhexidine, clean the tagging site in the ear and the applicator end of the tagging equipment.**

   The tag should go between the upper and lower cartilage ribs in the ear. If you were to imagine the ear split into thirds from right to left, the tag would go in the middle third. Figure 9-3 shows the correct location.

4. **Place the two halves of the tag on the applicator.**

   Be sure the stud (or male) portion of the tag is pushed firmly down on the applicator pin. Secure the female part of the tag (with the writing) with the clips on the tagger. If the two parts are mounted properly, the stud should line up with the hole in the tag.

   Numerous tag applicators and ear tag manufacturers exist. Not all taggers and tags are interchangeable though. Be sure your tags fit and work with your applicator.

5. **Place the tagger with the female part of the tag on the front side of the ear and the male part of the tag on the back side of the ear.**

   Be sure the applicator is properly oriented so the tag is placed in the correct location on the ear.

6. **Firmly and quickly close the tagger and then release.**

   You need to squeeze hard to get the tag through the skin. If you don't have a lot of hand strength, you may want to squeeze the tagger with both hands.

7. **Check that the tag is placed securely and comfortably.**

   The female part of the tag should swivel easily.

8. **Record the data on the tag along with a physical description of the animal.**

9. **Keep an eye on the tag site for several days until it heals.**

   In rare cases, the hole in the ear may become irritated and infected. If infection occurs, clean the area daily with chlorhexidine until it heals. If the infection gets worse, consult with your vet about treatment options.

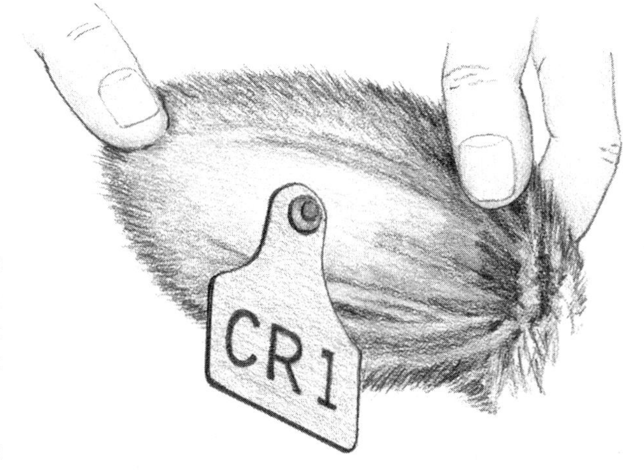

Figure 9-3: Correct placement for an ear tag.

Treating your cattle's wounds

Your cattle may suffer from cuts, scrapes, or puncture wounds sometimes. Minor injuries usually don't warrant treatment because the stress of handling is often worse than the wound. However, you should keep a close watch to make sure no infection sets in. If your cattle are tame and easy to work with, or if the injury is fairly large or especially dirty, go ahead and treat it.

Here's the general process for cleaning and treating wounds:

1. **Use sterile saline or water to flush any dirt or debris from the wound.**

   You can irrigate the area with a large syringe (without the needle).

2. **Clean the wound with a mild skin cleaner like chlorhexidine.**

3. **After the bleeding has stopped, cover the area with a water-based antibiotic ointment.**

   Be sure to use an ointment approved for animal use.

If the wound continues to bleed, place several gauze pads over the area and hold them firmly in place. Wounds that continue to bleed after several minutes of direct pressure may require stitching by a vet.

Chapter 10

Addressing Common Cattle Ailments

In This Chapter

- Identifying and controlling respiratory diseases
- Addressing clostridial illness
- Managing reproductive diseases
- Handling other infectious conditions
- Dealing with internal and external parasites
- Avoiding feed-related ailments

Cattle are hale and hardy creatures, but they always face the possibility of health problems. Luckily, if you're familiar with common ailments and the ways to prevent and treat them, you'll feel more confident in your ability to properly care for your cattle.

Don't let the potential difficulties make you hesitant about raising cattle, however. To offer a little perspective, consider this: During the last ten years, we have cared for more than 1,000 head of cattle, and we can count on less than our combined 20 fingers the number of times we've had to treat any of the problems covered in this chapter. By taking the proactive, preventive approach, you can usually keep your cattle healthy.

In this chapter, we take a look at some of the different health issues you may encounter in your adventures of raising beef cattle, including ways to prevent and treat those issues.

Understanding Respiratory Diseases

All classes of cattle can experience respiratory diseases. These conditions are costly in terms of the medications used, the labor needed to treat the diseases, and the decreased production. And, of course, death is a possibility (and a costly one at that).

Here are a few of the possible respiratory infections you may encounter:

- **Upper respiratory tract infections:** With upper respiratory tract infections, cattle experience inflammation of the nose, throat, and trachea. The animals may have a cough, nasal discharge, fever, and decreased appetite.
- **Diphtheria:** Infection of the voice box (larynx) is called *diphtheria*. Infected animals are noisy during breathing. If airflow becomes too restricted, they can die.
- **Pneumonia:** *Pneumonia* is an infection of the lower respiratory system and often results from the spread of an upper respiratory tract infection. This condition is serious. With this disease, you see many of the same signs as with an upper respiratory tract infection. However, you may also see more severe signs like rapid or labored breathing, no interest in eating, and lethargic behavior.

As you can see, no part of the respiratory system is immune from disease. And no single cause of respiratory disease exists. However, one or more of three main factors are usually involved: stress, viruses, or bacteria.

For effective control and treatment of respiratory diseases, make sure you do the following:

- **Minimize the stress your cattle experience.** When cattle are under stress, their immune systems have a harder time fighting off infection. Their well-being and performance suffer, too. Stress can come from the following factors:
  - Extreme temperatures (both hot and cold)
  - Excessive moisture in the form of mud or wet hair coats
  - Mixing cattle from different groups or sources (In such situations, cattle are exposed to new germs and fight to reestablish a pecking order.)
  - Dust, which can irritate the respiratory system and make it more prone to infection
  - Injury or surgery, such as dehorning
  - Problems with food or water (The animals may not be getting enough to eat or drink, or they may be experiencing a change in their diet.)
  - Situations causing anxiety, such as being transported or weaned

  As a cattle producer, always work to decrease the stress your cattle experience. Keep a close watch on your cattle during stressful periods and for a few weeks after them.

- **Develop and implement a proper vaccination plan.** Cattle often carry viral and bacterial agents in their systems but only exhibit mild clinical signs. However, if the cattle are stressed while infected or are challenged with multiple agents at the same time, a serious problem can take hold. So in addition to minimizing stress, vaccinate your cattle for the common infectious agents listed later in this chapter.

  Livestock health product catalogs are full of different types of vaccinations that you can order for your cattle. Visit with your local vet and the local beef extension specialist to learn about the specific immunizations that work for the viral and bacterial strains found in your particular geography. (To find your local cooperative extension service, go to `www.csrees.usda.gov/Extension/USA-text.html`.)

  When giving vaccinations, always read and follow the manufacturer's directions regarding the dose, route of administration, and timing. For instance, some vaccines require two doses to be fully effective. And you want to give the shots enough time to stimulate immunity before cattle are exposed to disease or stress. Chapter 9 covers the recommended vaccination schedule as well as directions for administering a shot.

  The vaccines for infectious bovine rhinotracheitis (IBR), bovine virus diarrhea (BVD), bovine respiratory syncytial virus (BRSV), and parainfluenza-3 (PI-3), which are described in the following sections, are often combined into a single product.

- **Provide supportive care if disease does occur.** Like the viral infection that causes a human cold, you can do little to treat viral agents once an animal is infected. The best you can do is offer supportive care. *Supportive care* includes comfortable, quarantined shelter, plenty of fluids, highly palatable feed, and pain or anti-inflammatory medicine if recommended by your vet.

- **Use antibiotic therapies properly.** Work with your vet to select an appropriate medicine, and begin the treatment as soon as possible in the disease process. Be sure to follow all package instructions, and continue the therapy for as long as directed (don't stop as soon as the animal appears normal). If the animal doesn't respond to the initial treatment, talk with your vet about changing therapies.

The following sections provide information specific to the most common respiratory diseases seen in cattle. These diseases can attack all ages of cattle.

IBR, BVD, BRSV, PI-3, pasteurella, and H. somnus (all described in the following sections) can contribute to bovine respiratory disease (BRD) complex. The viruses often impair the proper functioning of the upper respiratory tract, causing it to be unable to clean the air and mucus of disease-causing bacteria. The bacteria can then enter the lung and damage the tissue, thus impairing breathing ability.

Infectious bovine rhinotracheitis (IBR), or red nose

The viral agent (bovine herpesvirus 1) that causes infectious bovine rhinotracheitis, or IBR, is spread through the air and is most common in the fall and winter. It attacks mucous membranes that line the trachea, the female reproductive tract, and the eye.

In combination with a bacterial infection, the IBR virus can cause shipping fever or pneumonia. (*Shipping fever* is the respiratory sickness often seen in calves seven to ten days after transportation and after commingling with groups of calves from diverse backgrounds.)

Cattle infected with IBR have the typical signs of respiratory illness we mention earlier. In addition, they may have eye or vaginal discharge. One unique sign of the disease is the presentation of a red nose. Increased blood supply to the muzzle gives the nose an unusually red appearance. As a result, the disease is often referred to as *red nose.*

Although IBR is most often thought of as a respiratory problem, the virus can also infect a pregnant cow and cause an abortion 20 to 45 days later. It can also cause infections in the eye.

Immunizations can protect against IBR infection and are available in killed and modified live forms. Be sure to use killed vaccines on pregnant cows and/or the calves nursing pregnant cows. (You can read more about killed vaccines in Chapter 9.)

Bovine virus diarrhea (BVD)

Bovine virus diarrhea (BVD) is a family of viruses that negatively impacts not only respiratory function but also digestive, immune, and reproductive function. The virus spreads through contact with feces and secretions from the nose and mouth. It can also be transmitted through the air or from mother to fetus during pregnancy.

Around 95 percent of cattle infected with the BVD virus show no signs of disease. However, the BVD infection decreases the animal's ability to fight off other infections, so sickness may occur when the animal is challenged with a second disease-causing agent.

In acute cases where the BVD virus is the primary causative agent, infected cattle may exhibit high fevers, yellow discharge from the nose and eyes, sores on the muzzle and mouth, and diarrhea with mucus and blood. Chronic cases show the same signs plus bald spots due to hair loss and lameness due to inflammation around the hooves. Infected fetuses may be aborted, but some survive and are persistently infected.

Persistently infected calves are never able to rid their bodies of the virus and repeatedly shed the virus and infect other cattle. If you think you may have persistently infected cattle, talk with your vet about performing special virus identification tests.

No treatment for BVD exists. The only tool you have is preventive action via immunizations. Because there are two strains of BVD, be sure to consult with your vet to find out whether you need to vaccinate for only Type 1 or both Type 1 and Type 2 BVD. Use only killed vaccines on your pregnant cows and their nursing calves. (Check out Chapter 9 for more on using killed vaccines.)

Bovine respiratory syncytial virus (BRSV)

The bovine respiratory syncytial virus (BRSV) affects the lining of the respiratory tract so the animal becomes more susceptible to further infection. The disease is thought to be spread via secretions from the nose and mouth and is found mainly in cattle 4 weeks to 4 months of age (although it can be seen in older cattle, too).

It comes on quickly with varying degrees of severity. In mild cases, the calves have a dry cough with little to no nasal discharge and increased breathing rate. In both the mild and severe forms, the animal has a fever. As the disease worsens, the animal starts to breathe through the mouth and fluid collects in the lungs. Nasal swabs of potentially infected animals can be tested at a diagnostic lab to confirm the presence of the virus.

To help decrease the occurrence of BRSV, make sure calves consume plenty of colostrum (see Chapters 5 and 12 for more on colostrum). If you keep your cattle in a barn, avoid penning newborn calves with older calves to reduce the potential for disease transmission. The antibodies in colostrum protect calves from BRSV for a limited time, so be sure to vaccinate your calves against the virus as well.

Parainfluenza-3 (PI-3)

On its own, parainfluenza-3 (PI-3) doesn't cause much trouble, but because it makes cells of the upper respiratory tract more prone to further infection, it can become more severe in combination with bacterial agents. This viral agent's signs include cough, nasal discharge, and increased respiratory rate and breath sounds. Because PI-3 is common among cattle, be sure to vaccinate against it.

Pasteurella

Here are the two main pasteurella agents: *Mannheimia haemolytica* and *Pasteurella multocida.* These bacterial agents are widespread and can be major contributors to the development of pneumonia when combined with a virus and stress. The bacteria normally inhabit the nasal passages, but when cattle are stressed or under attack from viral respiratory diseases, they have reduced ability to clear the bacteria from their lungs.

The first signs of pasteurella infection are often slight depression and reduced appetite. As the disease progresses, the animal may refuse to eat, have a droopy head and ears, and have nasal discharge that changes from thin and clear to thick and yellow. Coughing and breathing become painful for the animal.

Vaccines to help prevent pasteurella infection are available. And if an infection does occur, the bacteria are susceptible to antibiotic treatment.

Haemophilus somnus (H. somnus)

Haemophilus somnus (*H. somnus*), in combination with infectious bovine rhinotracheitis (IBR), often causes shipping fever in recently transported cattle. It may also cause encephalitis (brain infection or fever), reproductive problems, arthritis, and heart muscle infections. Problems due to *H. somnus* most often occur one to four weeks after different groups of cattle have been brought together. Some cattle may exhibit signs of brain damage, such as blindness, staggering, or convulsions, and others may develop pneumonia.

Your vet can help you decide which antibiotic will be most effective in treating *H. somnus*. However, act fast to initiate treatment, because without proper treatment, death will likely occur. Vaccinations are available for *H. somnus*. The efficacy of these immunizations isn't clear, but because they don't seem to cause any harm, vaccinating your cattle is a wise move.

Closing In on Clostridial Diseases

Clostridial diseases are caused by organisms in the *Clostridium* class of bacteria. More than 60 different types of *Clostridium* bacteria exist, but not all cause disease. These bacteria are found in the soil almost everywhere livestock have lived, and they can lie dormant for years. In fact, they're naturally occurring in the intestinal tracts of livestock.

These diseases are usually spread when cattle come into contact with *clostridial spores* (reproductive structures that protect the bacteria from harsh conditions) on the ground. The spores are spread from infected animals or in times of heavy floods when large amounts of water expose previously buried spores. So animals that die of clostridial diseases should be burned or buried deeply.

When the spores enter the body, the bacteria grow and release toxins into the animal. Clostridial diseases often have a rapid onset and are difficult to treat. Therefore, preventive vaccines are critical to preventing death. These vaccines are commonly referred to as seven-way or eight-way vaccines because they immunize against seven or eight strains of *Clostridium* with one shot. Read the vaccine directions to determine which bacteria your cattle will be protected from.

Table 10-1 shows the different disease-causing clostridial bacteria, the name of the disease, the description of the disease, and the method of treatment or prevention.

Table 10-1 Common Clostridial Diseases

Bacteria	*Disease*	*Description*	*Treatment & Prevention*
C. chauvoei	Blackleg	Animals die within 12 to 48 hours after ingesting spores. Dead animals have swelling of infected muscle tissue that turns dark in color (hence the name blackleg).	Treatment with penicillin is rarely effective. A vaccine is available, however.
C. tetani	Tetanus	Enters the body during injury. Symptoms include severe muscle tremors, difficulty chewing food, and a stance like a "sawhorse."	You can treat with penicillin, antitoxin, and supportive care. A vaccine is available.

(continued)

Table 10-1 *(continued)*

C. septicum	Malignant edema	Generally fatal. Develops when a wound becomes infected with bacteria. Fluid accumulates in the tissue, causing edema.	No treatment available. Generally fatal, very prompt treatment with penicillin may be effective. A vaccine is available, however.
C. hemolyticum	Red water disease	Occurs in animals with liver damage (usually due to liver flukes). Damaged red blood cells discolor the urine with a reddish cast.	Early treatment with penicillin or broad-spectrum antibiotics is critical. Vaccinate for and control liver flukes. You can vaccinate for the bacteria, but not all clostridial vaccines immunize against this disease.
C. perfringens type C *C. sordellii*	Enterotoxemia	Usually seen in calves less than a month old that have had a sudden increase in food intake.	Provide intravenous fluids, electrolytes, antitoxins, and antibiotics. Vaccinate cows and keep your calves' diets consistent.
C. novyi type B	Black disease	Usually affects cattle on a high grain ration and causes liver damage. Animals die quickly. Liver will be gray to black in color.	Vaccinate for prevention. Death usually occurs before treatment can begin.
C. perfringens type D	Overeating disease	Found in older calves on a high-grain diet. Symptoms include decreased appetite, weakness, and nervousness. Death may occur rapidly.	Treat with intravenous fluid, antibiotics, and antitoxins. Vaccinate for prevention.

Dealing with Reproductive Diseases

Diseases of the reproductive system in cattle are sneaky. They often develop and spread slowly, so by the time you're aware of their presence, several animals may already be infected. Often, the animals, especially bulls, don't ever appear ill. However, these carriers pose a threat to the herd because they continue to spread disease. If you have a cow/calf herd, these diseases can really hurt your profits. They can also result in open cows (not pregnant), aborted calves, and poor growth and milking performance.

To keep your cattle reproductively sound, make sure you're familiar with the common diseases and work with your vet to create a preventive vaccination program. (Chapter 9 provides basic guidelines you can modify with your vet to fit your particular concerns.) Also, select new herd additions only from reputable sources, and then practice proper quarantine procedures as discussed in Chapter 7.

The four main reproductive diseases are brucellosis, leptospirosis, vibriosis, and trichomoniasis. We cover each of these diseases in detail in the following sections.

Brucellosis

Brucellosis, or *Bang's disease,* an infection of the bovine reproductive system, is caused by bacteria and can be found in other livestock species, pets, and humans. The bacteria can survive for several weeks in the placenta, birthing fluid, or an aborted fetus. Milk can also serve as a disease carrier and transmit the bacteria to calves and people.

Because people can be infected with brucellosis, wear disposable gloves when assisting cows during delivery.

You can't determine that cattle are infected just by their appearance. Although some cows with brucellosis abort, not all do. You may also see reduced weight gain and lower milk production. You can diagnose the disease using blood tests or tissue cultures.

State and federal governments have worked for many years to eradicate brucellosis. States are classified as brucellosis-free, Class A, Class B, or Class C. Brucellosis-free states haven't had any cases in the previous 12 months. Class C states have had more cases than A or B states.

Check with your vet about the class status of your state, and ask around at the locations where you purchase cattle. Brucellosis status can change at any time and impact your ability to transport cattle across state lines (not to mention increase your chance of bringing the disease to your herd).

Even though an individual bovine or a group has been tested and declared free of infection doesn't mean they're 100 percent certain to be disease-free. They could be in the incubation stage of the disease. To be safe, quarantine any new additions to the breeding herd and retest them in 45 to 120 days.

Heifer calves that you'll retain as breeding stock should be vaccinated between 4 and 12 months of age. This immunization must be done by a licensed vet. Heifers that have been vaccinated receive an official ear tag and tattoo. These females are referred to as having received their *calfhood vaccination* or having been *calfhood vaccinated.*

Leptospirosis

Leptospirosis, or *lepto* as it's often called, is caused by bacteria and occurs in livestock, wild animals, and humans. Abortions, especially in the third trimester, along with blood-tinged urine or milk are associated with the disease. It's spread through infected urine or aborted fetuses. Calves can be afflicted with the acute form of lepto that's characterized by a high fever and alternating bouts of diarrhea and constipation. A definitive diagnosis of this disease requires a lab test of serum.

Check with your vet to learn about the common strains of lepto in your area. You want to be sure to use a vaccine that confers immunity to these strains. Vaccination can protect against lepto for about a year.

Vibriosis

Vibriosis (*vibrio* or *campylobacteriosis*) is a venereal disease in cattle caused by the *Campylobacter* bacterium. Bulls pass the disease to cows during mating. Infected cows may abort their pregnancies or may fail to conceive.

Because several diseases affect the reproductive system like vibrio does, your vet will probably recommend culturing fluid or secretions from adults or tissue from infected fetuses to confirm the diagnosis.

To prevent vibrio, only add uninfected heifers into your herd. Vaccinate purchased and home-raised heifers twice before breeding. After that, vaccinate your cow herd and stud bulls annually.

Trichomoniasis

Another venereal disease of cattle is *trichomoniasis.* It's caused by *protozoans* (single-celled organisms) and is spread to cows by infected bulls. Infertility in the form of abortions and failure to rebreed quickly are signs of the disease.

Vaccinate breeding females in your herd against the disease. If they're already infected, treat the resulting uterine infection with medicine recommended by your vet. To give the body time to rid itself of infection, don't attempt to breed treated females for 90 days. If you're using bulls that aren't virgins, culture their semen to ensure they aren't infected.

Contending with Other Infectious Diseases

Infectious diseases can cause obvious problems like the loss of an eye, but they also can have more subtle impacts in terms of animal comfort and performance. For the well-being of your cattle and your bank account, be aware of the diseases in this section and work to keep them to a minimum.

Pinkeye

Pinkeye is a highly contagious eye infection caused by bacteria. It causes painful irritation of the eyeball and eyelid and, if untreated, may lead to blindness.

Pinkeye is often precipitated by eye irritation due to flies, dust, harsh sunlight, or tall grasses with seed heads and pollen. Face flies spread the illness as they travel from animal to animal. Calves are generally more susceptible to it than older cattle. Calves with pinkeye may weigh 8 to 10 percent less at weaning time than calves without the disease.

The symptoms of pinkeye initially are watering from the eye along with squinting and avoiding the sunlight. The disease can progress quickly with ulcers forming on the cornea in as little as two to three days. Check out Figure 10-1 to see a calf exhibiting the beginning stages of pinkeye as demonstrated by the small gray ulcer on the black part of the eye.

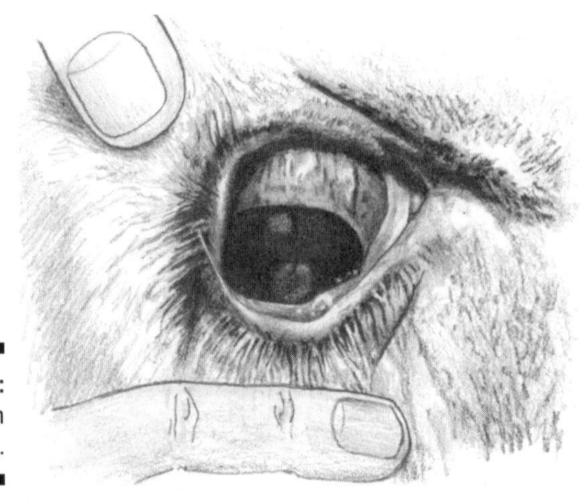

Figure 10-1: A calf with pinkeye.

Prevention of pinkeye is better for your cattle than treatment. Control face flies with insecticides (see the later section "Looking Out for External Parasites"). Also, if possible, keep pastures clipped or grazed to prevent plant pollen and seeds from entering and irritating the eye. Vaccines are available and should be given just prior to fly season. (Flip to Chapter 9 for more information on vaccinations and fly season.) Although vaccines may not always reduce the incidence of pinkeye, they do help decrease the severity of the disease.

When treatment is necessary, act early for best results. If you can handle your cattle easily in a low-stress manner, you can apply frequent doses of topical antibiotic on the mucous membranes around the eye. If you can't handle your cattle repeatedly, you can opt for an injection of antibiotic and corticosteroid. Check with your vet to determine the best antibiotics to use.

For best treatment results, cover the eye with a patch to protect it from further irritation. Be sure to leave the bottom side of the patch unsealed so moisture can drain away from the eye.

Johne's disease

Johne's (pronounced "Yo-nees") disease is spread by contact with the feces or saliva of an infected animal. It's usually transmitted from an infected mother to her calf soon after birth.

This chronic disease is caused by infectious bacteria. The cells of the small intestine absorb these organisms. Part of the body's immune response to remove the bacteria involves a thickening of the intestinal lining, which leads to poor nutrient absorption and diarrhea. The cattle still eat and feel fine, but they begin to lose weight. Sometimes you may never realize your animals have a problem because it can take up to two years for signs of infection to occur. Over time you begin to see a decrease in milk production and the weight of your infected cattle. They may also have lower feed efficiency.

Because no treatment or vaccine for Johne's exists, the key to controlling it is good management practices. Diagnostic tests, such as fecal cultures, can detect the presence of infection. If you know you have infected animals, make sure other cattle don't ingest their manure via a dirty udder while nursing or via contaminated feed or water. Keep the calving area clean of manure and well-bedded. If possible, cull the infected animals.

Tuberculosis

Bovine tuberculosis (TB) is caused by a bacterium called *Mycobacterium*. Different species of this bacterium can cause infection in all warm-blooded animals. The disease is most commonly spread by breathing in the airborne bacteria; it can also be contracted by ingesting feces-contaminated feed or water or when a calf nurses an infected mother.

Cattle with TB develop growths (tubercles) on their internal organs. With time, the cattle begin coughing, having nasal discharge, losing weight, and becoming weak.

TB is a rare occurrence in the U.S. beef and dairy cattle population; however, it's a serious disease because it isn't curable. All infected livestock must be destroyed. The presence of TB in the live animal is detected by a skin test. The carcasses of animals butchered at all state and federally-inspected slaughter plants are checked for tubercles.

To reduce the chance of TB infecting your herd, only buy cattle from a TB-free herd, keep new purchases isolated, and retest for TB in 60 days. Additionally, limit the contact between your cattle and wildlife, and protect your livestock's feed from wildlife contamination.

Scours

Scours is a diarrhea in young calves caused by infectious microorganisms like bacteria, viruses, and protozoan parasites. We discuss the prevention and treatment of scours in Chapter 12.

Looking Out for External Parasites

External parasites like flies, grubs, lice, ringworm, and mites are bad for your cattle in many different ways, including the following:

- The wounds and skin irritation caused by these parasites can make your animals uncomfortable and irritated.
- These pests spread diseases, and they make animals more prone to additional health problems.
- Cattle dealing with external parasites are less productive; they have slower growth and reduced milk production and reproductive ability.

Sanitation is key to controlling external parasites. Keep pest-breeding sites to a minimum by eliminating areas of wet manure, dirty straw, or decaying feed. Work to keep high-traffic areas around feeders and pens well-drained and clean. Avoid keeping cattle in confined quarters to help reduce the spread of lice or mites.

Chapter 9 describes the external pests you may encounter and helps you decide when your cattle may need treatment. In the following sections, we take a look at the various methods of controlling these parasites.

Parasite control products often have broad coverage, making them effective against grubs, lice, mites, and many internal parasites. Be sure to read and follow the label directions so you know exactly what you are and aren't treating.

Flies

Flies are challenging to control because they're so mobile and in contact with the livestock in short amounts of time. A variety of flies attack cattle, including horn flies, stable flies, horse flies, deer flies, face flies, and house flies. For best results, you need to apply treatments at the right times and in a consistent fashion.

Some of the options for dealing with flies include insecticide ear tags, contact control, and feed-through fly control. We talk about each of these techniques in the following sections.

Trying out insecticide ear tags

Insecticide ear tags look like most of the tags used for identification purposes except they contain insecticide that kills flies. The flies get a lethal dose by coming into contact with the insecticide that has rubbed onto or around the animal's head. The insecticide is also spread to the body during grooming or mingling with other cattle.

The two main classes of insecticides used on these tags are pyrethroid and organophosphate. To prevent the flies from building resistance to the tags, switch between these two insecticides every year or two. Be sure to use the correct number of tags (1 or 2) as directed by the manufacturer. Also use the correct type (adult or calf) of tag. The tags have a limited duration of effectiveness of about 12 to 15 weeks, so use fresh tags, not leftovers from last year. Attach tags when you see about 50 flies on each side of the animal. See Chapter 9 for proper tagging techniques.

Don't handle these tags with your bare hands; wear protective gloves. Thoroughly wash hands and arms with soap and water when taking a break and after tagging. Wash your tagger, too.

Remove the tags at the end of fly season to lower future resistance. You can purchase a tag remover from your vet or farm supply store. This plastic tool is about 6 to 8 inches with an encased metal blade at one end and a handle at the other end. Because it's flat and sharp, you can lay it on the backside of the ear between the tag and the ear and with one swift tug slice the tag in two without hurting the animal.

Opting for contact control

Contact control methods work through direct application onto the cattle so that when flies land on the cattle, they come into contact with the insecticide. The products come in the form of sprays, pour-ons, or self-applicators. They provide effective control for a limited duration, so you need to be able to reapply frequently. Start using these methods as soon as the fly season starts.

Here's the lowdown on each of the contact control methods available:

- **Sprays:** With these products, mist your animal in the areas where flies congregate (head, neck, back, and legs). They work nicely for cattle that you work with on a daily basis because they give good protection if you're close enough to the cattle to apply them properly. Be sure to pack fly spray with your cattle show gear. It's hard to get your calf to pose for the judge when it's being attacked by biting, buzzing flies.

  When we transport cattle on a livestock trailer during fly season, we always spray them right before loading. Doing so is a good way to leave the flies without any bovine hosts and to start the cattle off in their new home fairly fly-free. You also can use this method if you're rotating between pastures that are spaced far apart (at least half a mile).

- **Pour-ons:** These products are applied directly to the animal's shoulders and back. Many pour-ons have a wide spectrum of control and are effective against flies, lice, mites, and grubs. Be sure to check the weather forecast before applying, however; some products require the cattle to stay dry for up to 24 hours after application.
- **Self-applicators:** Self-applicator products are usually cloth or rubber flaps or bags with oil or dust insecticide on them. You place the applicators where the cattle come into contact with them daily (such as across the gate opening to a new pasture or water trough). Check the applicators often to make sure the insecticide hasn't run out. If your applicators apply insecticide in dust form, protect them from rain so the dust doesn't cake up.

If you're using both ear tags and contact control, use two different types of insecticide for broader coverage.

Using feed-through fly control

Feed-through fly control products are oral larvicides that your cattle ingest as part of their feed or mineral supplementation and then pass out with the manure. The chemicals prevent the larvae from developing into pesky adult flies. Because these products work by disrupting the fly life cycle, they don't get rid of mature flies that come over from your neighbor's herd, so you'll need to use a second control method if you have unwelcome visitors. Use feed-through products from just before the start of fly season until the first hard frost in the fall.

Grubs

Grubs are the larvae of heel flies that grow and develop inside your animal along the throat or backbone. The buzzing of the heel fly really bothers cattle, and the larvae make cattle lose weight and damage their hides. You can use pour-on, spray, or injection treatments (the treatment for grubs can be given in a liquid shot form) to kill grubs.

Be sure to use grub products *before* the larvae have migrated to the throat or back to prevent choking or paralysis (see Chapter 9 for details on treatment timing).

If you think your cattle may have grubs but you didn't treat for them before they reached the throat or back, use care when selecting a product to treat for other parasites, such as internal worms or lice. You need to avoid products that affect grubs as well (to avoid choking or paralysis).

Lice

Two main types of lice affect cattle: chewing and sucking lice. The *chewing louse* feeds on hair and excretions from the skin, and the *sucking louse* punctures the skin and feeds on blood. Because lice infestations irritate cattle, reduce their appetite, and make them more susceptible to disease, you want to keep these pests under control. Lice are mainly a problem in the late fall, winter, and early spring.

Cattle with lice often have hair loss, and they rub against objects in an attempt to relive the itching and irritation. However, other problems can cause the same signs, so examine the animal to be sure. If you part the hair in areas where lice are often found, such as the head, neck, shoulders, or the base of the tail, you can see the lice or their nits attached to the hair. The *nits* are eggs, and they hatch one to two weeks after being laid. In about another two weeks, they start laying their own eggs.

Several of the products and application methods that are effective against other parasites also work on lice. You can use sprays, self-applicators, and pour-ons. Injections are another option. Insecticide fly tags don't work to control lice.

Many insecticides don't affect nits, so you need to give two treatments 14 to 18 days apart to get the adults and their subsequently hatched offspring.

Ringworm

Despite its name, ringworm isn't a worm. It's a skin and scalp disease caused by several different fungi, and it's transmitted by direct contact with infected skin or hair. Ringworm can also be passed via equipment used on infected animals (like hair brushes) or through the use of the same barn. It occurs most often in bovines on the face and neck.

You can catch ringworm from your cattle. If your cattle have ringworm, always wear gloves when touching them or working with things the cattle may have contacted.

Initially, the affected areas are small patches of rough hair that are slightly raised. As the fungus infects the hair follicle, the hair falls out, leaving distinctive circular-shaped gray sores. As the disease spreads, bald patches up to 3 inches in diameter may form. Take a look at Figure 10-2 to see a cow with ringworm.

Because an infection usually clears up on its own and the disease causes little, if any, permanent damage, treatment isn't necessary. However, vets won't issue health certificates to cattle with ringworm, which means those animals can't participate in shows or exhibitions.

If you want to show your cattle or are worried about the ringworm spreading to other cattle (or yourself), treat the condition. While wearing gloves, apply a topical medication like 2-percent iodine solution or thiabendazole paste directly to the lesions. If the lesions have scab-like coverings, remove the crusts so the medication can reach the fungus. Be sure to collect the crusts and double-wrap them in plastic bags for disposal to prevent contamination of the environment. Treat the lesions at least twice, three to five days apart. Disinfect the barn and grooming equipment used by your cattle. A mixture of one part bleach to three parts water is an effective solution to use.

Figure 10-2: A cow with ringworm.

Mites

Mites are tiny eight-legged pests that can live on cattle or under their skin. They damage the skin and increase the risk of bacterial infections. The diseases associated with mites are mange and scabies.

Because mites can burrow under the skin, injectable products or pour-ons that work with systemic (whole body) activity are the best treatment options. Like lice, an initial treatment with a follow-up two to three weeks later is needed to kill adults and their newly hatched offspring.

Controlling Internal Parasites

Internal parasites include stomach worms, lung worms, liver flukes, coccidian, intestinal worms, and tapeworms. Internal parasites can cause your cattle to underperform. Animals infested with these parasites grow slower, have lower reproductive and feed efficiency, and are more prone to other diseases than cattle without parasites.

To best control internal parasites, administer the dewormer (anthelmintic) when cattle are infected, when the parasite is developing and most susceptible to treatment, and when conditions are conducive to parasitic transmission. For more on the timing of treatment, head to Chapter 9.

Anthelmintics come in several forms:

- **Pastes:** These are thick formulations that you deposit into the back of the mouth with a special gun that resembles a caulking gun.
- **Suspensions:** You squirt these liquid forms into the animal's mouth.
- **Boluses:** You place these large, pill-shaped treatments beyond the base of the tongue with a *balling gun* (a strong plastic or metal device that holds the bolus and then releases it when the plunger is depressed).
- **Crumbles:** These are small, edible pieces of anthelmintics that you mix with the feed. They eliminate the need to catch and restrain animals for treatment. However, to make sure cattle get the correct amount of dewormer, they need to have the feed available at all times.
- **Injectables:** You administer these via a shot. Check out Chapter 9 for shot-giving tips.
- **Pour-ons:** You squirt these formulations on the animal's back.

When giving dewormers orally with a paste, suspension, or bolus, restrain the animal in a head gate so you can be sure it receives the proper dose.

The active ingredients in anthelmintics have names like albendazole, fenbendazole, ivermectin, and levamisole. To ensure that the parasites don't develop resistance to your treatments, alternate between the different types available. Even if the brand name is different, be sure that the active ingredients aren't the same so that over time you don't run into resistance problems.

Anthelmintics have *withdrawal times,* which means it can take anywhere from a few days to several weeks from the time of treatment until the cattle can be safely and legally slaughtered for human consumption.

When using pesticides, always read and follow label directions. If you get chemicals on your skin or clothing, wash skin thoroughly and remove clothing. Store pesticides in their original containers and keep them out of reach of kids and animals.

Preventing Feed-Related Problems

Feed-related health problems can be due to feedstuffs that aren't wholesome or to rapid changes in the diet. These problems are usually preventable, so in the following sections, we discuss the potential issues to avoid.

Moldy feed

Mold and the toxins it produces (mycotoxins) can be found in feed that has been grown, harvested, or stored in less than optimal conditions. These situations can range from cool, wet growing seasons to hot, humid weather. Or wet hay may have been baled or products may have been stored in leaky grain bins. When cattle ingest moldy feed, it can cause problems with their respiratory, digestive, and reproductive systems.

Moldy feed may have a musty odor or other unpleasant smell. You may be able to see the mold, whether it is black powder on corn or gray mold on hay. Sometimes, however, you can't see or smell the mold or its mycotoxins. So if your cattle experience widespread health problems, send feed samples to a lab for testing. If the results come back positive for mold, stop feeding your cattle the contaminated feed and consult with your vet regarding possible treatment.

Poisonous plants

Because poisonous plants vary so much by region of the country, your beef extension agent is your best source of information about the poisonous plants of specific concern in your area. (To find your local cooperative extension service, go to `www.csrees.usda.gov/Extension/USA-text.html`.) These plants can cause a wide range of physical problems, so make sure you can recognize both the plants and the symptoms. Remove the offending plants from your pastures either by pulling them out or treating them with an herbicide.

Cattle avoid eating most poisonous plants because they often taste bad. If you ensure that your animals have adequate amounts of hay and pasture to graze, they naturally avoid the poisonous plants.

Founder

Founder, or laminitis, is a condition that often occurs when an animal has experienced an abrupt change in diet, especially when going from a high-roughage diet to a high-grain diet.

With this condition, the starch from the grain is broken down in the rumen and produces large amounts of lactic acid. The increase in acid can kill rumen microbes and release toxins into the bloodstream. These changes in the blood alter the blood flow to the hoof, causing it to grow abnormally. The misshapen hooves make walking uncomfortable and difficult for the cattle. As a result, they're less likely to eat properly, and they may experience reduced weight gain and performance.

Founder can be prevented by gradually changing the diet of your cattle over a few weeks to give the rumen bacteria time to adapt. Also, be sure to keep feedstuffs stored out of the cattle's reach, so even if the animals somehow escape their pens or pastures they can't overindulge. Similarly, when grazing the crop aftermath of harvested grain fields, allow your livestock limited access to the fields in case there was a grain spill or lots of downed crops in the field. After the cattle become acclimated to the crop aftermath, you can give them free access.

Bloat

Bloat is the buildup of gas in the rumen. This gas is formed through the normal digestive process and is usually released by belching. But during bloat, either something is blocking the throat (as occurs in *gassy bloat*) or a stable foam forms on the top of the rumen contents blocking the release of gas (as in *frothy bloat*). Bloat is painful for cattle and can lead to death. Animals with bloat have a swollen left abdomen, and they'll be in pain and distress.

The following list gives you the information you need on both types of bloat:

- **Frothy bloat:** This type of bloat is the most common, and it usually occurs when animals are grazing lush spring and fall pastures with a high legume content. For more about legumes, refer to Chapter 6. The legumes make the rumen contents sticky, so gas bubbles have a difficult time forming and escaping the body. To treat frothy bloat, use a stomach tube to put an antifoaming medicine into the rumen.
- **Gassy bloat:** This type of bloat occurs less frequently. You can pass a stomach tube down the animal's throat to relieve the gas buildup. After the gas is gone, you then have to ask your vet to find what was causing the obstruction and to remove it.

It's better to prevent bloat rather than treat it. To do so, introduce cattle to high-legume lush pastures gradually when their appetites are already fairly satiated. Also continue to give your cattle access to high-fiber feeds like grass hay or straw to keep the rumen contents in balance.

Sweet clover poisoning

Sweet clover poisoning occurs when clover plants have been damaged or molded. This spoilage releases a toxic compound that prevents normal blood clotting. Symptoms include stiffness, dull behavior, swelling, pale mucous membranes, and blood in feces or urine.

The problem is found most in animals that are consuming clover hay; it's rarely seen in grazing situations. Don't feed spoiled or moldy clover hay, if possible. If you must feed toxic hay, alternate it weekly with good alfalfa hay. Animals with poisoning may need a blood transfusion or a shot of vitamin K.

Chapter 11

Breeding Cows and Caring for Pregnant Females

In This Chapter

- Getting your cow ready for pregnancy
- Managing your bull
- Breeding your cows without owning a bull
- Preparing for the birth
- Observing the stages of labor
- Taking care of problems during pregnancy

A robust calf nursing at the side of a well-conditioned mother cow brings a smile to the face of any cattle herdsman. To achieve such a peaceful and bucolic scene takes diligent care and management, but when you wean a group of healthy calves, all the work is worthwhile.

Before you have a weaned calf to sell, show, or grow to maturity, you first need to prepare your cow for pregnancy and then breed her. In this chapter, we talk about these tasks along with how to help deliver your cow's calf. We also discuss how to handle problems that may occur during pregnancy or after delivery.

Preparing Your Cow for a Successful Pregnancy

A little preparation goes a long way toward helping your cow have a healthy pregnancy and a vigorous newborn calf. Keeping your cow in good physical condition, giving her age-appropriate care, and performing the needed preventive healthcare all help make breeding and caring for her a rewarding experience.

In the following sections, we show you how to assess a cow's physical condition, how to prepare younger and older cows for pregnancy, what reproductive vaccines to administer, and how to determine a due date.

Assessing body condition scores

An underfed pregnant female can experience the following undesirable consequences:

- She will give up too many of her own body reserves to nourish her pregnancy. As a result, she'll have more trouble giving birth.
- The quality of her colostrum will be reduced, causing the calf to receive less antibody protection.
- The cow won't produce as much total milk for her calf.
- The cow will be slower to rebreed, which means she won't produce a calf every 365 days.

To ensure that their cattle are getting the nutrients they need, herdsmen have developed a nine-point system to assign body condition scores. *Body condition scores* (BCS) are numbers used to quantify the relative fatness, or body condition, of your mother cows before, during, and after pregnancy. Thin cows are assigned a lower number, and fat cows are given a higher number. Table 11-1 lists the physical characteristics of the different BCS, and Figure 11-1 shows examples of cows with various BCS.

Table 11-1 Body Condition Scores and Associated Visual Signs

Body Condition Score	*Physical Characteristics*
1	Very emaciated and weak; bones distinctly visible. Cattle with this score are rare and usually have diseases or parasites.
2	Very emaciated but not weak. The tailhead and ribs protrude. Muscles are atrophied in the shoulder and hindquarter.
3	Very thin with no fat over the ribs or in the brisket. The backbone is visible, and muscle depletion is present in the hindquarters.
4	Slightly thin condition with the spine and three to five ribs showing. Muscle isn't depleted. Some fat is visible over the ribs and hips.
5	Moderate condition; the last two ribs aren't fat covered, but the spine is smooth and the backbones don't jut out.

Body Condition Score	*Physical Characteristics*
6	A well-conditioned, smooth appearance throughout the body. Some fat is present in the brisket and over the tailhead. The back is rounded, and fat is visible over the ribs and pin bones.
7	Very good flesh; brisket is full of fat and tailhead has fat deposits around it. The ribs are smooth, and the body has a soft, rounded look with hip bones slightly visible.
8	Fat and overconditioned. Excess fat makes the neck appear short and thick and the back look square. The bone structure is covered by fat.
9	Very obese and rarely seen.

At calving time, your cows should be a body condition score of 5 to 6, and first-calf heifers should be at 6 (or higher if they're calving during cold weather).

We recommend evaluating your cows' BCS at specific times of the year, such as the second trimester of pregnancy, before calving, and prior to breeding. Doing so allows you to make any necessary adjustments in the feeding program so your cows get the nutrients they need.

A one-level change in the BCS of a medium-sized cow that isn't pregnant constitutes a loss or gain of 75 to 100 pounds. For a pregnant cow, a one-level change in BCS is more like 175 to 200 pounds, with the extra 100 pounds due to the fetus and weight gains associated with pregnancy. So if during the second trimester your cow has a BCS of 4 and you want her to be a BCS of 5 at calving, she needs to gain 175 to 200 pounds.

On our farm, we pay closest attention to our pregnant cows' BCS during the middle of winter, when they're in their second trimester. The cows' nutritional needs are lowest during the second trimester, but the weather is the harshest and the feed sources are limited in quality and quantity. If the cows have a BCS of 4 or 5, we supplement their pastures with moderate-quality hay. Our cows calve in mid-May when grass is abundant and nutritious, so we know that they will rapidly gain weight and condition and be at a BCS 6 at calving time.

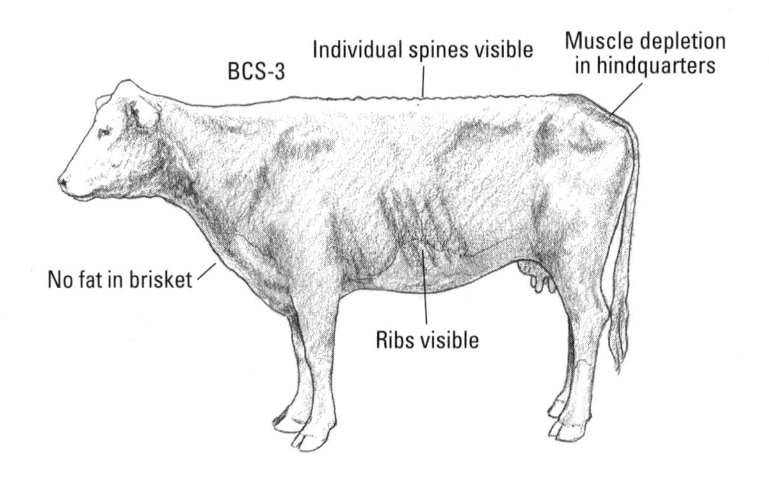

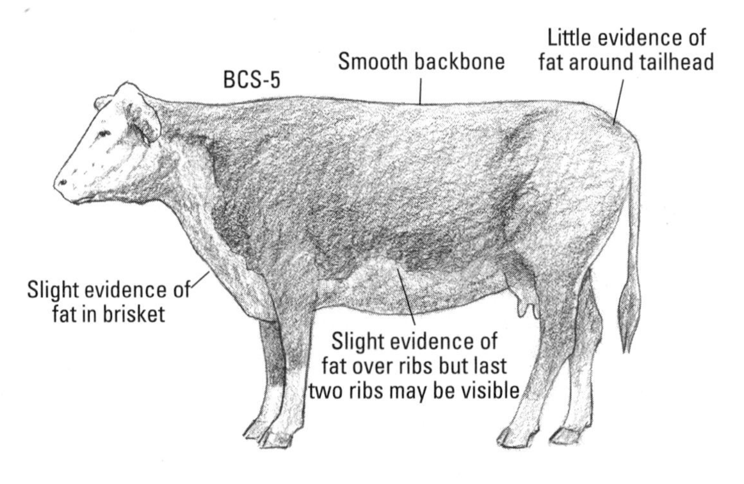

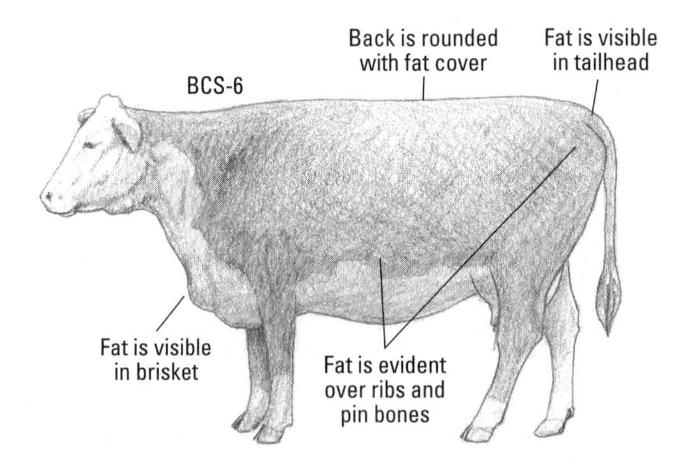

Figure 11-1: Examples of the appearance of different BCS.

Providing extra preparations for cows at either end of the age spectrum

Of course, all cows need to be cared for and prepared well for pregnancy, but cows who are 3 years and younger (generally first-calf heifers) or who are more than 9 to 10 years of age benefit from extra care and preparation for pregnancy. We explain what you need to do for these groups in the following sections.

Tending to your first-calf heifers

First-calf heifers, or pregnant females giving birth for the first time, are not only nurturing a developing pregnancy but are also still growing themselves, which puts a lot of extra stress on their bodies. (These cattle are also often called *replacement heifers.*) However, if you select your heifers carefully and provide them with additional nutrition, you can have them successfully give birth at around 2 years of age, rebreed to calve again in about a year, and still wean *growthy* calves (meaning the calf puts on at least an average amount of weight and height compared to calves of a similar age). Properly nurturing your first-calf heifer goes a long way toward helping her become a profitable member of the herd.

Here are some practices you can employ to increase your chances of success with first-calf heifers:

- **Use a rigorous selection process for the heifers you buy or keep from your own herd.** Look for females whose dams rebreed quickly, have good dispositions and maternal abilities, provide ample milk, and wean a healthy, growthy calf. The potential replacement heifers should reach puberty by 12 to 14 months and be at least 65 percent of their mature weight. If you can evaluate the pelvic area of the heifers, make sure the pelvic opening is big enough for easy birthing. (The pelvic area in 12-to-14-month-old heifers weighing around 600 pounds should be 150 to 160 square centimeters.)
- **Feed the heifers to provide for their proper growth and the development of the calf.** Use the following stages as guides:
  - **From weaning to breeding:** During this stage, most heifers need to gain 1.25 pounds per day. If heifers need to catch up on their growth to reach the 65 percent mature weight target, don't exceed a gain of 2 pounds per day. Heifers that need to grow more than that are probably not good candidates for bringing into the mother cow herd. To achieve a growth rate of 1.25 pounds per day, heifers need around 12 to 15 pounds of dry matter a day. Unless the pasture provided is high in both energy and protein, you should supplement their pasture with concentrate.

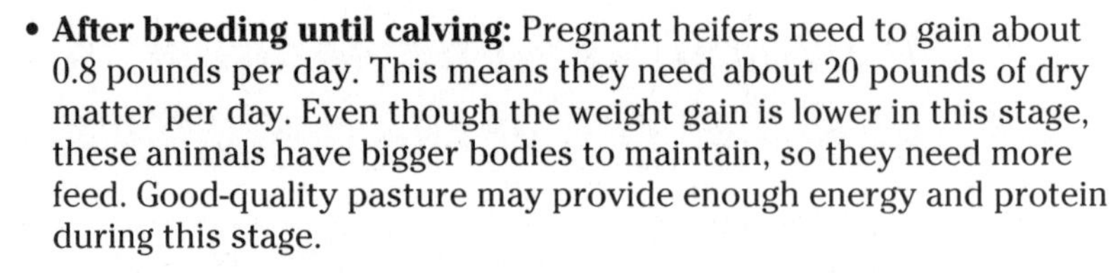

- **After breeding until calving:** Pregnant heifers need to gain about 0.8 pounds per day. This means they need about 20 pounds of dry matter per day. Even though the weight gain is lower in this stage, these animals have bigger bodies to maintain, so they need more feed. Good-quality pasture may provide enough energy and protein during this stage.

  Improving the condition of a heifer as she approaches calving is difficult, and it's even harder to get her to gain weight immediately post-calving. So supplement her ration with concentrate as necessary to maintain a body condition score (BCS) of 6 to 7 pre-calving. (Check out the earlier section "Assessing body condition scores" for more on BCS.)

- **After calving:** The first-calf heifer continues to grow and won't reach her mature weight until the time of her second calving. This is especially true of the larger, later-maturing breeds. These cows should be in a BCS of 5 to 6. They may be able to maintain this status while eating with the main cow herd, or you may find that they benefit from being managed with the first-calf heifers.

Make sure your females have access to a high-quality mineral and vitamin supplement at all times. For more information on specific nutrient requirements and feed options, see Chapter 5.

- **Breed first-calf heifers three to four weeks before the main cow herd.** This breeding schedule gives the heifers more time to rebreed after having their first calf while still staying in synchrony with the older cows. If they have to breed at the same time as the older females, try to keep their breeding season at 35 to 45 days. Doing so helps you retain heifers that have better reproductive efficiency and that are more fertile over the long run.

- **Use a bull with desirable birth weight and calving ease expected progeny difference (EPD).** Calving ease EPD predicts the ease with which a bull's calves are born. This EPD is largely influenced by calf birth weight, but it also takes into account other factors, such as calf body shape. In general, the bigger the calf, the more likely *dystocia* (calving difficulty) is to occur.

Because it takes a heifer more time to recover from a difficult birth and because she may be slower to rebreed, be sure to use a bull that's in at least the top 25 percent of his breed for low birth weights and calving ease. For breeds that aren't associated with calving ease, select bulls that are in the top 10 percent. (Chapter 3 discusses which breeds are recognized as good choices for mating to first-calf heifers, and Chapter 7 explains EPDs.)

Supporting older cows

You may see the production of your older cows decrease over time. They may have lower conception rates, or they may raise lighter-weight calves. This decrease in production may be due to their decreased foraging abilities. As they get older, they may be less likely to travel about the pasture, or their teeth may not be as sound. Both scenarios make them less efficient grazers.

If your older cows still have functional udders that produce enough milk to raise calves to a weight you deem appropriate, it can be beneficial to manage them with your first-calf heifers or 3-year-old cows. They will have access to better feed and face less competition for it.

With good, consistent care and preventive health maintenance, your cows can remain productive well into their teens.

Administering reproductive vaccines

Diseases affecting the reproductive system can have a significant negative impact on the health and profitability of your herd. They can keep cows from breeding, cause abortions, and hinder growth and milking ability. In addition to bringing in new animals only from healthy, well-managed sources and doing an adequate quarantine, you should work with your vet to develop a preventive vaccination schedule.

Here are vaccination guidelines for many diseases that affect the reproductive abilities of your cattle (for details on each of these diseases, see Chapter 10):

- **Brucellosis:** Vaccinate heifer calves that you intend to keep for your breeding stock herd between 4 to 12 months of age. A licensed vet must administer the vaccine and identify the heifers that have been vaccinated with an official ear tag and tattoo.
- **Leptospirosis:** Vaccinate heifers and cows at least annually to protect against this disease. If your herd has experienced a problem with leptospirosis in the past, you may need to vaccinate every six months. Check with your vet to learn about the common strains of leptospirosis in your area and to confirm that you're using the appropriate vaccine.
- **Vibriosis:** Vaccinate heifers twice before breeding, and give your mature cows an annual booster.
- **Trichomoniasis:** This disease is more common in the western part of the United States. Check with your vet to determine whether you need to vaccinate against this disease.

- **Bovine virus diarrhea (BVD):** A family of viruses causes this disease, which not only impacts the reproductive system but the respiratory, digestive, and immune systems, too. Two strains of BVD exist, Type 1 and Type 2, so be sure your annual vaccine protects against the strains your cattle may encounter.
- **Infectious bovine rhinotracheitis (IBR):** This disease attacks the female reproductive tract as well as the trachea and eye. Administer a yearly vaccination in heifers and cows.

If you're vaccinating bred cows and other cattle (such as herd bulls, yearlings, or nursing calves) in contact with bred cows, be sure the vaccines are safe for use in pregnant animals.

Your cows (and other bovines) also need preventive healthcare in areas not related to reproduction. For information on a total wellness program for all cattle, see Chapter 9.

Determining a due date after conception

To give a cow appropriate care during the different stages of her pregnancy, or *gestation,* you need to know when she's due to give birth, or *calve.* The length of gestation in cattle is around 285 days. So if a cow is bred and conceived on January 1, you can expect her to calve around October 12.

It isn't uncommon for a cow to give birth seven to ten days before or after her expected due date. So be prepared early and also be prepared to be patient.

Following a vaccination schedule

Because each breeding program is unique, we can't recommend a cookie-cutter immunization routine. However, you can work with your vet to develop a program that addresses the potential problem areas in your herd and your geography. After creating your plan, you need to find a convenient way to consistently follow through. If, for example, you move your cows to new pastures after calving and before breeding season, it may make sense for you to do the herd's annual health maintenance then. The correct vaccination schedule is the one that offers the best protection for your cattle and that you can accomplish in an efficient, low-stress manner.

The stage of pregnancy (or lack of a pregnancy) can be determined by *palpation*. Palpation is done by inserting a gloved arm into the animal's rectum and feeling the reproductive tract for changes associated with pregnancy. An experienced palpator can detect pregnancy as early as 30 days after breeding. However, you can be more confident in an examination done 45 or more days after a cow's last exposure to a bull or artificial insemination.

Because the fetus can be fragile in the early stages of pregnancy and determining the gestational age can be more difficult in the later stages of pregnancy, it's best to have a highly experienced person, such as a vet, pregnancy check your cows.

Tending to the Male Side of the Reproductive Equation

One of the most important decisions you make when raising cattle is the selection of your herd bull. That bull is responsible for half of the genetic performance of your calf crop. And if you decide you want to grow your herd by keeping the heifer calves you've raised, the bull's genetic influence will impact your cow herd for years to come because his daughters will start producing calves of their own.

Depending on your situation, your breeding season may be as short as 45 days or as long as 90 days. Plan on a yearling bull servicing 15 to 20 cows. A mature bull in good condition and with excellent libido can handle up to 35 cows.

Because this guy is so important to your herd, in this section, we discuss what to look for in a bull, how to make sure he will perform his duties, and what kind of care he requires.

Selecting the right bull for your cows

You want to pick a bull that helps you achieve the goals you've set for your cattle-raising enterprise. The bull you choose should complement the strengths of your cows and help improve their weaknesses.

Consider what you need from your bull by asking yourself these questions:

- **Will he be used on heifers?** If so, calving ease and birth weight are of paramount importance. Flip to the earlier section "Tending to your first-calf heifers" for a discussion on these topics.
- **Will you keep all his daughters to expand your cow herd?** If you plan to keep his daughters, choose a bull that's known to produce cows with strong maternal traits like fertility, milking ability, udder conformation, and mothering ability.
- **Are you selling his offspring as feeder calves?** If so, a bull known for siring calves with big weaning weights may be the ticket.
- **Do you market your calves as beef direct to the consumer?** If so, purchase a bull that excels in carcass traits. Chapter 3 discusses carcass merit in more detail.

Regardless of your goals for your bull, you should assess four main areas before buying:

- **Reproductive soundness:** For a bull to be of any value, he must be fertile. The best way to determine reproductive soundness is through a breeding soundness exam. We discuss the details of the exam in the upcoming section "Understanding the breeding soundness exam." For now, just remember that a bull that passes a breeding soundness exam is physically capable of breeding and *settling* your females (making them conceive) at the time of the exam.
- **Structural fitness:** To efficiently and consistently breed cows, a bull must have a sound body structure. The following should be true of him:
  - He should move freely without pain or discomfort.
  - His body must be able to withstand the rigors of mounting cows.
  - His hooves should be free of disease or injuries.
  - He should have no swelling of the legs and joints.
  - He needs good vision so he can see the cows that other cattle are mounting. A bull with poor vision can also be dangerous to handle.
  - His teeth and mouth should be in good condition so he can eat easily and comfortably.
- **Visual appraisal:** Look at your potential herd sires thoroughly. A visual appraisal is the best way to evaluate them for traits like disposition, muscling, color, body condition score (5 to 6 is a good score at the start of the breeding season), horn status, and visual appeal. Chapter 7 goes into more detail about what to look for before you buy.
- **Performance ability:** Check out the bull's own performance in areas like birth, weaning, and yearling weight. Ask whether ultrasound or feed

efficiency data is available. Of great importance are the bull's expected progeny differences, or EPDs. The EPD information is useful in determining how your animal may perform as a parent in areas like growth, maternal ability, and carcass value. As you evaluate potential sires, you can check out their calving ease EPD so you can be more likely to select a bull whose calves are born easily.

Understanding the breeding soundness exam

A *breeding soundness exam* is an evaluation of a bull's ability to get cows pregnant. The exam is done by a vet or by experienced beef extension personnel and consists of the following evaluations:

- **Physical evaluation:** The exam conductor looks at all the points related to the bull's structural fitness (refer to the earlier section "Selecting the right bull for your cows" for details).
- **Reproductive organ assessment:** The evaluator inspects all the reproductive organs, noting certain characteristics for each. Here's what he assesses:
  - **Internal reproductive organs:** Examined by rectal palpation for inflammation or scar tissue.
  - **The scrotum and its contents:** Checked for abnormal growths, injuries, and frostbite.
  - **Testicles:** Inspected to determine whether they're firm and even in size.
  - **Penis and sheath:** Checked for sores, warts, or adhesions, which may make breeding painful or even impossible. The penis is also assessed to determine whether it fully extends without any signs of discomfort.
- **Measurement of scrotal size:** The circumference of the scrotum corresponds to daily sperm production and is an easily repeatable measure, which makes it a valuable indicator of semen production. Bulls with bigger testicles and hence larger scrotal size generally reach puberty earlier and sire sons with bigger testicles and daughters that become reproductively mature at an earlier age. Scrotal size varies with breed, age, and external temperature, but the minimum acceptable scrotal circumference for a yearling bull is 30 centimeters; the minimum for a mature bull is 34 centimeters.
- **Semen check:** The evaluator collects semen from a bull either when he mounts a cow or via *electroejaculation* (a probe is inserted into the rectum, and it emits an electric pulse causing the bull to ejaculate). After

collecting the semen, the evaluator looks at it under a microscope. He inspects the sperm cell concentration as well as the motility (movement) and morphology (shape) of the sperm cells. A good semen sample has more than 90 percent motile sperm and at least 75 percent that are normal in shape.

There is no practical way to measure a bull's *libido* (mating ability) other than to watch him mounting cows. Take the time to observe your bull with the cows to make sure he's finding the females in estrus (heat) and successfully mating with them. A bull that's hesitant to mount cows may have an injury that makes breeding painful. It's a very rare bull that doesn't have the desire to mate. But for whatever reason, if your bull isn't servicing the cows, you'll need to make other arrangements to get them pregnant.

A breeding soundness exam is just a snapshot of the bull's reproductive ability. A bull that's deemed fertile today may suffer an injury or disease that renders him infertile later. However, if you're buying a bull, it's still worthwhile to have the exam done so you don't buy a "dud" when you want a "stud."

Caring for your bull

Because your bull is an important contributor to the success of your cattle herd, you need to take care of his nutritional and health needs so he's fit and ready to go to work when breeding season starts.

Bull calves that are growing from weaning time to their first breeding season need to gain about 2.5 pounds per day to ensure proper development. During this time, an average bull grows from around 600 pounds to 1,000 pounds, and his dry matter intake increases from 15 pounds to 23 pounds. Growth can continue until the bull is 3 years old, so feed your bull enough to keep him around a body condition score of 6. (Refer to the earlier section "Assessing body condition scores" for details.) Older bulls on a maintenance diet consume from 25 to 31 pounds of dry matter daily, depending on their size. (Chapter 5 gives specific ideas for possible feedstuffs and feed plans.) If your bull has been on a high-concentrate diet for test-station (see Chapter 7 for more on test stations) or exhibition purposes, you need to gradually adjust his diet to one similar to his breeding-season diet.

In most cases, your bull should be with the cows for a specific breeding season, not all year long. By limiting the duration of time your cows can get pregnant (a typical breeding season is from 60 to 90 days), your herd will be in the same stage of production (pregnant, lactating, rebreeding) so you can care for their specific feed and health needs more precisely. In addition, if your bull is with the cows and their calves for most of the year, he can prematurely breed any young heifers that become sexually mature earlier than expected. Another bonus is a *tight calf crop* (one where the calves are similar in age). It's easier to properly care for calves that are close in age and have similar dietary and health needs.

When your bull isn't with the cows, house him in a dry, clean area with protection from the weather. Try to pasture him in an area where he can exercise. We like to put our bull's water trough at one end of the pasture and the feed bunk at the other end so he doesn't get too lazy and stand around all day.

Your bull can be on the same vaccination schedule as the mature cow herd. He needs the following vaccines:

- **Reproductive:** Leptospirosis, vibriosis, and possibly trichomoniasis
- **Respiratory:** Infectious bovine rhinotracheitis (IBR), bovine virus diarrhea (BVD), parainfluenza-3 (PI-3), and bovine respiratory syncytial virus (BRSV)

Deworm your bull in the spring and fall. And be aggressive with his fly-control program. With all their saliva-flying, head-tossing ways, bulls attract a large number of flies (which then cause diseases and become irritating). Chapter 10 provides information on fly control.

Getting Cows Pregnant without Owning a Bull

A bull can be a big investment both in money and management. Bulls also sometimes go looking for love when you least want them to. We take a look at two options you have for getting your cows bred without the hassle of bull ownership.

Opting for artificial insemination

Artificial insemination (AI) is the practice of using collected semen to breed a cow instead of having the bull mate the cow naturally. About two-thirds of the nation's dairy cows are bred using AI. However, only about 5 to 10 percent of the country's beef cows are bred with AI. In the following sections, we discuss the pros and cons of AI and give you an idea of what's involved so you can determine whether you want to give it a try.

The pros and cons of AI

The benefits of AI include the following:

- **You gain access to high-quality sires.** The biggest advantage to artificial insemination compared to natural service is the increased access to highly proven, top-performing AI sires. Using such bulls can increase weaning weights, improve carcass value, and produce valuable replacement heifers more so than average herd bulls.
- **You don't need a bull.** Breeding your cows with AI can eliminate the need for a bull. And not needing a bull can make some herdsmen quite happy.
- **You'll likely spend less than with a bull.** When all the costs of AI are totaled, you may find it more cost-effective than buying a bull, especially when you only have a few cows.

Here are some drawbacks associated with AI:

- **You may face lower conception rates.** The conception rate for AI is usually lower than for natural service. So keep in mind that after two attempts at AI, most producers put the cow with a bull for natural service.
- **You need more equipment, labor, and skill.** Compared to letting the bull do all the work, AI requires more effort, time, and equipment. The upcoming section gives you the rundown on what you need.

Deciding whether AI is right for your cows

Artificially inseminating your cows is a lot of work, but it can be done successfully. You can hire an AI technician to do the work for you, or you can do it yourself. In this section, we provide you with an overview of what's involved so you can decide whether it's a viable option for your herd. First, keep in mind the things you need:

- **Proper equipment and supplies:** You need to have proper working facilities with a corral to hold the cattle and a head gate and chute to safely restrain the female. This setup reduces stress for man and beast, and it also helps prevent injury.

  Required supplies include bull semen, a tank filled with liquid nitrogen to store the semen, tweezers, a thermos with warm water, a thermometer, an insemination gun, paper towels, a semen straw cutter (or sharp scissors), a sheath for the insemination gun, long plastic gloves (also good for palpation and during calving), a clock, and a record book.
- **Time:** You need to have the time to observe your cattle twice a day to determine which cows are standing in heat. *Standing heat* or *estrus*

is when the cow stands immobile when mounted by another bovine. Standing heat happens every 18 to 24 days and lasts for around 12 hours. About 50 percent of cows show estrus between 6 a.m. and noon or from 6 p.m. to midnight. Only 10 percent stand in heat in the afternoon, so early morning and late evening checks are ideal. And make sure they aren't eating as you observe; otherwise, they'll be more interested in eating than they are in each other. Also keep in mind that the best conception rates occur when cows are bred at the end of their heat. This timing allows the sperm to travel through the reproductive tract and meet up with the egg when it's ovulated and ready for fertilization, which is about 12 hours after heat ends.

You can put large stripes of livestock marking chalk on each of your cows from their pin bones to tail. When they're mounted by other animals, the chalk becomes smeared or completely rubbed off. This indicator tells you to watch that individual for signs of estrus.

If you see a cow in heat during your morning observation, artificially inseminate her that evening. If you have a cow standing in the evening, breed her early the next morning.

- **Skill:** If you decide you want to artificially inseminate your cows yourself, you need training in the areas of insemination technique, semen handling, and reproductive management. Training classes are offered by several semen suppliers and should give both hands-on and classroom training. In these classes, you learn how to accurately and sanitarily place semen into the reproductive tract, how to handle semen so it stays fertile, and how to detect heat in your cows. You also should get direction on how to remove semen from the tank and get a chance to practice with the actual equipment.

Plenty of in-depth resources are available. Large semen suppliers, such as American Breeders Service (`www.absglobal.com/beef`) or Select Sires (`www.selectsires.com/index.html`), provide equipment, education, and, of course, semen to cattle breeders across the country.

Semen is stored in a plastic straw, and several straws are held by one metal cane. The cane is suspended in a canister inside a tank filled with liquid nitrogen. The liquid nitrogen keeps the contents at –320 degrees Fahrenheit. Semen won't be viable if it gets above –112 degrees, so you must keep your tank filled with at least several inches of liquid nitrogen.

Use your record book to keep accurate records of the location of each straw so you can quickly and accurately remove the desired straw while keeping the rest of the semen cold. Store your tank on a pallet in a room with adequate ventilation. Use caution not to breathe the nitrogen gas when opening the tank.

The following set of steps gives you some basic knowledge about the procedure so you know what you're getting into:

1. **After using tweezers or forceps to remove a semen straw from the tank, thaw it in warm water stored in a thermos.**

   Semen suppliers have specific guidelines for thawing their semen, but generally the straw needs to be immersed for around 40 seconds in 90- to 95-degree Fahrenheit water (use a thermometer to check the water temperature).

2. **Dry the thawed semen straw with a paper towel, and then give it a flick to get all the liquid contents to the end opposite of the crimped portion of the straw.**

3. **Insert the straw into the insemination gun, and then cut off its crimped end.**

4. **Put the plastic, sterile sheath over the gun, and then wrap the sheath, gun, and semen straw in a clean paper towel and tuck it within your clothing for transport to the cow.**

   Wrapping and tucking the equipment in your clothing helps keep the semen from getting too cold. Yes, the semen was stored at –320 degrees Fahrenheit, but after you thaw the semen, it must stay warm.

5. **Speak quietly to the cow as you approach her in the chute.**

   You want to speak to her so she's aware of your presence. But you need to speak quietly so you don't frighten her.

6. **Use paper towels to gently clean manure from the external surface of her rectum and vulva. Then insert your gloved arm into the rectum and the insemination gun into the vagina.**

   You need to feel through the walls of the rectum for the reproductive tract and insemination gun. Guide the gun through the vagina about 6 to 8 inches until it reaches the cervix. The cervix is made of dense connective tissue and muscle. (For details on cow anatomy, check out Chapter 2.) After the tip of the gun reaches the cervix, use your gloved hand to guide the cervix onto the gun. Use gentle and consistent pressure to direct the gun to the end of the cervix. For best conception rates, the semen needs to be deposited at the end of the cervix where it meets with the uterus. Slowly pushing the plunger on the gun moves the plug down the semen straw and forces the semen into the uterus. Figure 11-2 shows the proper positioning of the insemination gun so the semen is deposited in the uterus.

7. **Carefully remove the insemination gun and your gloved arm from the cow.**

8. **Record the cow identification, breeding date, and semen used.**

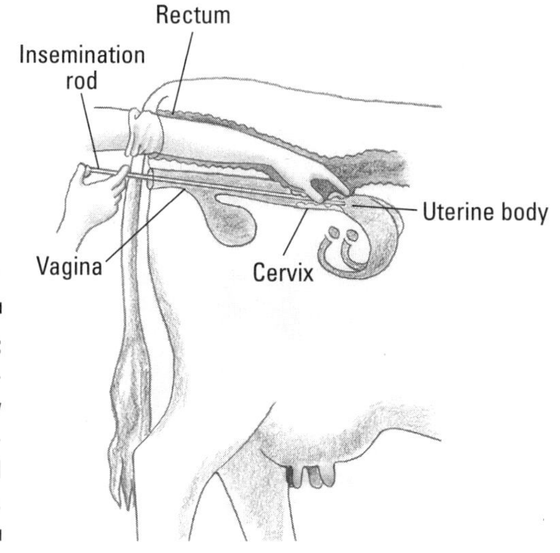

Figure 11-2: Cross-section view of cow during artificial insemination.

If you take more than 15 minutes from the time you remove the semen from the tank to depositing it in the cow, the semen probably won't be viable anymore. So remove all distractions and learn to work quickly yet carefully.

Leasing a bull

Leasing a bull provides a way to get your cows pregnant without all the drawbacks of owning a bull or doing AI. You don't have the hassle or expense of keeping a bull all year (you only need to keep him for the length of your breeding season), nor do you have the extra effort of checking for heat and performing AI.

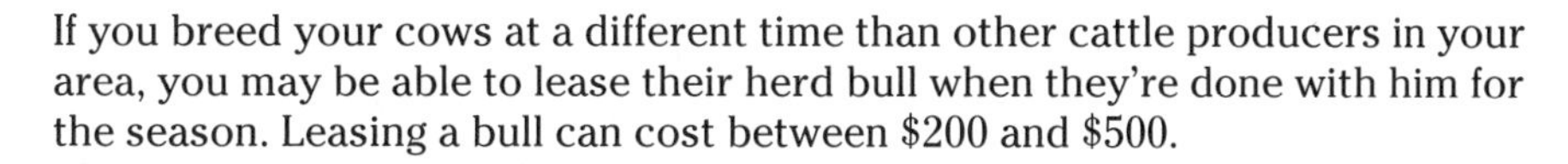

If you breed your cows at a different time than other cattle producers in your area, you may be able to lease their herd bull when they're done with him for the season. Leasing a bull can cost between $200 and $500.

Here are two big issues you should consider before leasing a bull:

- **You risk introducing disease into your cow herd.** Reproductive diseases can pose serious risk to your herd. (We discuss these diseases in Chapter 10.) Using a virgin bull can help reduce the risk, but he may still harbor disease (of course, the same may be true of a bull you buy).
- **It can be difficult to find a bull to lease.** Most breeders have concerns about disease transmission. They're also more interested in selling a bull outright than just loaning him out for a few months at a fraction

of the purchase price. If you can find a willing producer and a bull that you feel good about both genetically and health-wise, leasing is a viable option to consider.

Planning for the Big Delivery

After all the time and effort you devote to breeding and caring for your pregnant cow, it's important to stay diligent as calving time approaches. To help get the new calf and its mother off to a good start, be sure to have your facilities and supplies all ready to go. We show you how in the following sections.

Setting up the birthing spaces

Depending on your climate during birthing season, calving can take place on a well-drained pasture hillside or in a calving shelter. No matter the location of your calving facilities, however, they need to be clean and dry. Dirty and wet bedding is an invitation for the spread of disease. We show you how to prepare both types of spaces in the following sections.

Creating a calving pasture

An ideal calving pasture faces south or east to capture the sunlight and break the wind. To help control the spread of disease, leave the calving pasture empty for three to four months before calving. Ideally, each cow should have a ½-acre pasture space before calving and 2 acres afterward. Move the cows onto the pasture a few weeks before calving.

Keep these important reminders in the forefront of your mind; they'll help you keep your cows and calves healthy and disease-free:

- **Separate cows with older calves from pregnant cows.** As cows calve, either move the pairs to fresh pasture or leave the cows and calves in the pasture where birth occurred and move the cows that haven't given birth to fresh ground. Keeping these pairs separate helps stop the transmission of diseases from older calves to weaker newborns.
- **As cows give birth, work to keep them away from mud.** Mud makes cows and calves wet, and wet cattle need more energy to stay warm. Mud also serves as a place for disease-causing organisms to grow and spread. Chapter 4 reviews how to build a heavy use area protection (HUAP) zone to help minimize mud buildup.

- **Be sure to have shelter available in case of emergency.** If you have a cow that's having trouble calving or if a weak calf needs extra care, this shelter is important. Check out the next section to get the nitty-gritty on what your shelter needs.

Building a calving shelter

Producers whose cows calve in harsher weather should plan to have facilities that can house all the cows and calves that will be born over a two- or three-day period in case they experience a stretch of really bad weather. A shelter can be a simple three-sided building with pens that are around 10 x 12 feet.

The shelter needs a nonslip floor that's easy to clean. It should be well-lit and should have easy access to water and good ventilation without drafts. Make sure the pens have strong, easy-swinging gates so you can contain an upset mother cow swiftly and securely. The facility also needs a basic head gate that can be used to restrain the cow in case she needs help delivering or to hold her still as you help the calf nurse. (Chapter 4 provides information on equipment used to prepare your facilities.)

Unlike in your regular working facility where the head gate is attached to a stationary holding chute, swinging gates that can be moved or chained into position are better choices for a calving area. The moving gates allow the cow to have the freedom to lie down during delivery while keeping her comfortably restrained.

Gathering your calving supplies

Well before the first calf is due to arrive, you need to inventory and restock your calving supplies, including the following:

- **Basic items:** Be prepared with a flashlight with good batteries (in case you have to go checking the pasture at night for a wayward cow), paper towels or old towels, and a bucket for water so you can wash up after the delivery.
- **Calf feeding supplies:** For the calf to get off to the best start in life, it should consume only colostrum in the first six hours after birth. Some cows (especially first-calf heifers) may not produce adequate amounts of colostrum, so have an extra supply available. You can either reconstitute a dry powder or use frozen colostrum collected from a disease-free cow in your herd. Use a bottle or esophageal tube feeder to get this milk into the calf. For more on caring for the newborn calf, go to Chapter 12.
- **Mechanical calf puller:** For particularly difficult deliveries or when a cow isn't trying hard enough during labor, you may need to use a

mechanical calf puller. This contraption cups around the cow's rear end and has straps going over the cow's back to hold it in place. One end of your obstetric chains (explained later) is secured around the calf's legs and the other end of the chains is attached to a ratcheting mechanism that helps ease the calf out with more strength than what you can do by hand.

Practice extreme caution when using a mechanical puller. Don't pull too hard on the calf or cow. Also watch that the cow doesn't whip the puller around and hurt you.

- **Medications:** Here are a few of the medications you may need during labor and delivery:
  - **Uterine boluses:** During a difficult labor, cows have an increased chance of uterine infection due to the additional trauma and introduction of infectious organisms when a calf is pulled. Uterine *boluses* (very large antibiotic pills that release medicine as they dissolve) can be placed directly in the uterus to help fight infection.
  - **Broad-spectrum injectable antibiotics:** This medicine fights infections that may occur with a difficult delivery or retained placenta (see "Watching for postpartum actions" for more on these potential problems).
  - **Oxytocin:** Sometimes a cow hasn't received the needed hormonal signal to *let down,* or release her milk from the mammary glands so the calf can nurse. The hormone oxytocin can stimulate the release of the milk.

  Have clean needles and syringes available so you can administer these medications.
- **Obstetric chains with handles:** To assist a cow during a difficult delivery, you may need to help her by pulling the calf from the birth canal. An unborn calf can be strong and slippery, so the chains with stainless-steel handles (not twines or ropes that can't be sanitized) help you keep a firm hold on its legs. The proper placement of the obstetric chain around the leg involves putting a loop around the leg above the fetlock joint and a half-hitch knot below the fetlock joint. The connecting chain should be on the top side of the leg. If you're still having problems after trying obstetrical chains, try a mechanical puller, described earlier.
- **Obstetric sleeves and lubricant:** To prevent the spread of disease (between yourself and your cattle), use long plastic sleeves that cover your hands and arms. Lubricant makes it easier and less uncomfortable for the cow if you need to do an internal exam of the birthing canal.

Monitoring the Stages of Labor

Months and sometimes years of anticipation and hard work come to fruition as your first cow prepares to give birth. Most of the time the process goes smoothly, but you need to be on the lookout to avert potential difficulties before they become even bigger problems. To be able to properly care for your pregnant cows, you need to be able to identify the three stages of labor:

- Pre-delivery
- Active labor
- Postpartum

This section describes the three stages of labor, explains how a normal delivery should proceed, and advises you on what to do if issues arise.

Gearing up for pre-delivery

Pre-delivery, the first of the three stages, involves internal preparation of the cow and calf for delivery as the cervix begins to dilate. Signs of pre-delivery preparation include:

- The cow may show slight behavioral differences, such as signs of discomfort or a desire for isolation.
- She may lift her tail or switch her tail.
- She may have increased mucus discharge from her vagina.

Pre-delivery preparation normally lasts two to three hours in cows and four to five hours in heifers. However, longer amounts of time aren't abnormal. This stage ends when the water bag (pregnancy membranes) appears at the vulva.

Progressing into active labor

Active labor is the stage when the calf is born. The cow should have intense contractions to expel the calf. As delivery proceeds, the contractions become stronger and more frequent. The passage of the calf through the birth canal may take from one to three hours in cows and two to four hours in heifers. During this time, it's quite normal for the cow to repeatedly get up, lie down, and spin around. These movements help change the position of the calf in the birth canal and aid in delivery.

For the most part, cows and heifers can deliver their calves by themselves, but you should keep an eye on your animals while they're calving. You may need to assist in the delivery if it gets too difficult for the mother.

Dystocia (birthing difficulty) occurs in 10 to 50 percent of heifers and 3 to 5 percent of mature cows. Knowing when to assist is easier if you check your cows frequently. Usually a cow needs assistance if she hasn't made significant progress in one to two hours. Lack of significant progress can mean no signs of active labor one to two hours after the water bag is passed, or one to two hours of active labor without signs of the calf's feet and head emerging from the vagina and vulva.

Timing your assistance properly is critical. Helping too early can be harmful if the cervix isn't fully dilated. Waiting too long weakens the cow and calf and makes it more difficult for the calf to start nursing.

You should intervene sooner rather than later if you notice that the calf isn't in the normal (anterior) birth position. It's perfectly fine to call the vet for help if labor isn't progressing normally. However, you should know how to handle these situations on your own in case a vet is unavailable for immediate assistance. Here's a rundown of the most common birthing positions you may see in cattle:

- **Anterior position:** This is the most common birthing position. You generally won't need to assist with a delivery in which a calf is in this position unless the calf is large and has trouble fitting through the pelvis bones and birth canal; in such a case, you may need to use obstetrical chains or a mechanical puller. The calf's chest is on the bottom side of the cow's pelvis, and its head is resting between its front legs. The front legs should emerge first from the cow with the hooves facing down. The head follows along with rest of the front legs. The shoulders and rest of the body pass out of the cow next. The normal, or anterior, birthing position is shown in Figure 11-3.
- **Posterior position:** The calf enters the birth canal with the hind legs coming first. You can recognize this presentation because the hooves on the hind legs point upward instead of down. You'll also notice that bending the fetlock joint lifts the hoof up rather than lowering it down. With this position, the head emerges last, which puts the calf at risk of suffocation. This lack of oxygen can occur because the umbilical cord connecting the calf to the placenta may break while the calf's head is still inside its mother.

If you find a calf in this position, deliver it quickly by using traction (either pulling with your hands or obstetrical chains) on its hind legs. You may need to keep a hand on the calf's tail as it enters the birth canal so it doesn't damage the upper wall of the cow's vagina.

- **Anterior position with head retained:** The calf enters the birth canal in the anterior presentation but has the head turned back. Figure 11-4 illustrates this presentation.

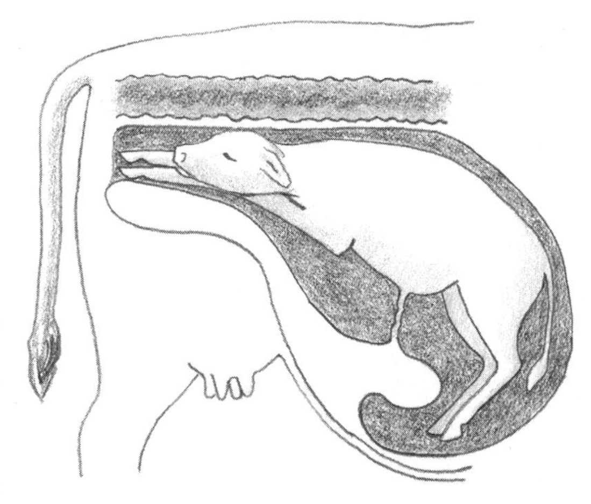

Figure 11-3: Normal calf birthing position.

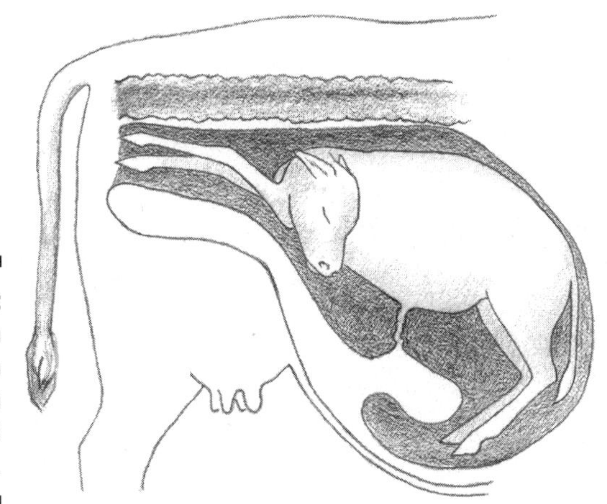

Figure 11-4: A calf in the birth canal with its head retained.

Don't let this position fool you into thinking the calf is coming backward. Determine whether the legs you see/feel are the front or hind legs using the fetlock joint methods we describe earlier. Grasp the muzzle of the calf and pull it around between the front legs. ***Warning:*** Don't pull hard on the jaw; it's fragile and may break.

- **Anterior position with retained limb(s):** Sometimes one or both of the front legs are turned back. Find the retained limb and grasp it above the fetlock. Bend the knee and push it toward the cow's spine. Meanwhile, work your hand down the leg until it is covering the hoof. After you have the hoof, carefully straighten the leg out into the birth canal. Bringing around a retained leg (or head) is easier if you push the calf out of the birth canal and back into the uterus where you have more room to work.
- **Breech position:** A calf in the breech position enters the birth canal butt first (see Figure 11-5). To deliver a calf from this position, push the calf back into the uterus. Grab the hind leg and work your way down to the hoof. Withdraw your arm as you bring the leg over the rim of the pelvis into the birth canal. Do the same procedure with the second leg. At this point, deliver the calf as described for a posterior presentation.

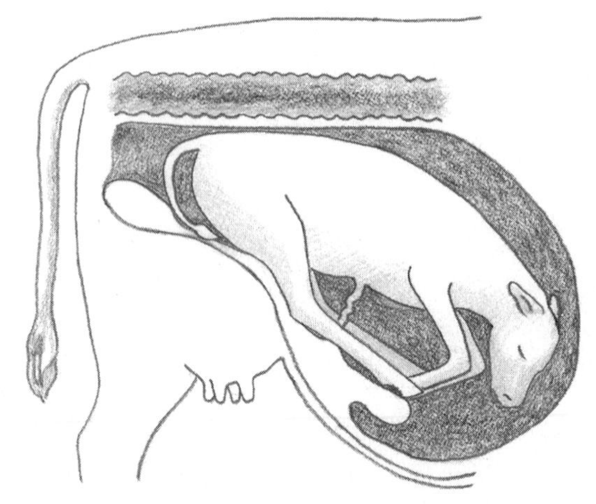

Figure 11-5: A calf in the breech presentation.

Excessive or improper pressure during delivery assistance can injure the mother cow or the calf.

When helping a cow or heifer deliver, remember that the calf needs to go through the birth canal in the shape of an arch. First it has to be lifted upward into the cervix. Then it goes straight through the cervix into the vagina. As it exits the vagina and vulva, pull in a slight downward direction to ease its passage out of the cow.

Seeing double

If you're helping deliver a calf and you can't seem to sort out the legs and head, consider the possibility of twins. One out of every 100 to 250 births in beef cattle is a twin. The legs of two different calves can enter the birth canal at the same time! You need to be sure the legs and head entering the birth canal all belong to one calf by feeling one leg and following it up to the chest and head and then back down the other leg. Once you have one calf (and just one, no parts of a second calf are allowed because there won't be enough room) deliver as described in this chapter. You should go ahead and help with the birth even if the calves are in a normal anterior presentation so that the calves don't have a chance to get tangled up again.

Watching for postpartum actions

Stage 3 of calving is the expulsion of the placenta. This expulsion should occur in no more than 8 to 12 hours after the cow gives birth. If the placental membranes aren't shed within 12 hours, the placenta is considered to be *retained.* This condition most often happens in first-calf heifers or after a difficult delivery. In the past, the placenta was manually removed, but new research has shown that it's better to treat the cow with antibiotics to prevent infection from the placenta and let the membranes slough off (and out) on their own. This sloughing should take no more than four to seven days. If the placenta hasn't passed by then, consult with your vet.

Managing Potential Issues during Pregnancy and after Delivery

If you've done the groundwork to properly prepare your cows for pregnancy, most gestations and deliveries go smoothly and without problems. However, it's always prudent to have a plan and supplies at the ready if issues do arise. Three difficulties you may encounter include abortions, milk fever, and paralysis. We cover all three in the following sections.

Coping with abortions

Abortion is the premature death and expulsion of a fetus from its mother. Cattle can abort at any time during their pregnancies. Early-term abortions may go unnoticed, because at three months the fetus is only about the size of a rat. A cow can lose her pregnancy and your first sign of a problem may be when she comes up open at pregnancy check time or when you observe her in heat. With later-term abortions, you may find the fetus or notice birthing fluids or blood around the cow's vagina.

An abortion rate of 1 to 2 percent may occur in healthy cow herds. Abortions can be due to a variety of causes, including the following:

- **Dietary problems:** Toxin-containing plants, such as ponderosa pine needles, locoweed, and broom snakeweed, can induce abortions in cattle. Also, improperly stored or harvested grains and forages can have dangerous levels of mold or nitrates that can cause abortions. Some drugs, such as modified live vaccines, can also cause your cattle to abort.
- **Genetic abnormalities:** Curly calf (arthrogryposis) and chromosomal defects are two types of genetic abnormalities that can induce abortions.
- **High fever associated with general body infection as well as specific reproductive infections:** Because diseases that cause abortion can be highly contagious, consult with your vet immediately if you encounter them. Doing so can help keep the problem from spreading. Be prepared to supply your vet with the following information:
    - Your vaccination protocols
    - Any movement of new cows into your herd
    - The management and origin of your bulls
    - The history of your cows with regard to any previous abortions, any problems with conception rate, and the affected cow's approximate stage of gestation

  Isolate aborting cows from the rest of the herd to help contain infection and make treatment of diseased individuals easier. Save fetuses and fresh placentas for testing. Don't wash or cut them. Wrap them in plastic, leak-proof bags. If the fetus is frozen, keep it frozen; if it's not frozen, refrigerate the specimen between 38 to 45 degrees Fahrenheit. A *necropsy,* an exam of the specimen, may offer clues to the cause of the abortion.

Some diseases that cause abortions in cattle can be transmitted to people. Always use caution and wear disposable plastic gloves when handling aborted fetuses and associated tissues or when working with the cow's reproductive system.

Being aware of milk fever

Milk fever (hypocalcemia) is a metabolic disorder that occurs most commonly in dairy cattle, but it can also affect older, high-milk-producing beef cattle. It happens when large amounts of calcium are moved from the blood of the cow to be used in making colostrum, but then the cow's body is unable to use reserves of calcium in the bone or absorb calcium from the digestive tract to replenish her blood supply. Most cases happen 48 to 72 hours before calving.

Symptoms of this disorder are muscle tremors, lack of appetite, and unsteady gait. The condition progresses as the cow becomes unable to rise and her body temperature falls.

Death can result if prompt treatment isn't given.

The recommended treatment for severe cases is intravenous (IV) injection of calcium solution. Oral administration of calcium may be an option in milder cases.

Work closely with your vet when using IV calcium because it can lead to fatal cardiac problems.

Proper dietary management can help reduce the incidence of milk fever. Your vet, beef extension specialist, or nutritional consultant can give you advice about adding anionic salts to your cattle's ration to enhance calcium resorption from the bone and absorption from the digestive tract. Also, avoid feeds high in potassium, because high levels of this mineral predispose animals to hypocalcemia.

Dealing with temporary paralysis

In the late stages of gestation or after a difficult delivery, cows may experience pressure, swelling, or trauma to the nerves that signal the hips or legs. This damage prevents the nerves from functioning normally, thus causing the cow to temporarily be unable to stand. If the cow is close to delivering, she and the calf may be okay without treatment. After the pressure from the fetus and associated pregnancy fluids and tissues is gone, the nerves may regain function. If they don't, you can administer medications to reduce swelling around the nerve. Similarly, if the damage occurs during delivery, administer medications to reduce swelling.

If the cow is unable to stand, hoist her up twice a day with a support strap under her belly. Doing so allows her to regain some strength and movement in her limbs without risk of falling and incurring further injury.

Preventing udder issues

A growing problem in beef herds is *mastitis,* or inflammation of the mammary gland. This disease, along with the ensuing infection, damages the udder and reduces milk production. Usually just one of the four quarters is affected, but the disease can spread to the entire udder as well. Although mastitis can occur at any stage of lactation, it most often happens in the first month of milking.

In acute cases, the udder is swollen and hot to the touch. Because the udder is so sensitive, the cow may kick at the calf when it tries to nurse. The cow also may have a fever and may not eat. In subclinical cases, the cow experiences inflammation and reduced milk production without the other more noticeable signs or symptoms.

Flies help spread the disease because their bites near the teat expose live tissue to disease-causing bacteria. (For tips on keeping flies at bay, head over to Chapter 10.) Calves can also spread the disease because they nurse from one quarter of the udder to the next. Only about 1 percent of mastitis is due to physical injury.

The best way to prevent mastitis is by maintaining cleanliness. For instance, keep mud to a minimum. And if the cows are confined, provide them with plenty of clean bedding. However, clean grass is even better than bedding.

If you suspect that your cows have mastitis, consult with your vet immediately. You need to be quick and aggressive with treatment to prevent permanent udder damage. Treatment may include systemic antibiotic therapy along with direct antibiotic treatment of the infected udders.

Chapter 12

Looking After Calves Young and Old

In This Chapter

- Tending to the newborn
- Managing problems in the early days
- Caring for the older calf

Calves are sturdy yet fragile, and cute but ornery. After 9½ months in its mother's womb, the calf makes its difficult entrance into the world. If everything goes according to nature's plan, your main duty is to stand back and watch the wondrous sight unfold. The cow licks her calf dry while it struggles to figure out how to coordinate its body and stand. In an amazingly short span of time, the calf takes its first teetering steps to its mother's udder for some rich colostrum.

In this chapter, we go over the basic animal husbandry duties involved with caring for a new calf. Even though calves are quite strong, their health can quickly take a turn for the worse. So we alert you to potential signs of trouble and tell you what to do if illness strikes. The needs of the older calf differ from those of the newborn, so we explain how to adjust your care as your calves mature. We also tackle ways to deal with dehorning, castration, and weaning.

Mother cows, especially right after calving, can be protective of their newborns. So be cautious while working around the cow and calf. Even tame, family pets may undergo a big personality change during calving.

Welcome, Baby: Caring for a New Bovine Arrival

We've had the privilege of watching the birth of many farm animal babies, and it's always a miracle. A newborn calf emerges with great effort on its mother's part as a big (usually in the range of 60 to 110 pounds), wet bundle. A cow with good maternal instinct usually takes care of all her calf's needs. However, you may want to perform some animal husbandry tasks or help your cows with some difficult situations. For instance, you should clean its navel and apply some sort of identification to the newest bundle of hooves. And you may need to help the calf start breathing and help it get the colostrum it needs. We explain all you need to know in the following sections.

Helping the calf with breathing difficulties

After the umbilical cord breaks during labor (see Chapter 11 for more on the birthing process), the calf should start breathing on its own. However, after a long or difficult labor, the calf may be sluggish and slow to breathe. Here are some tips for getting the process going:

- If the cow doesn't lick the placental membranes from the calf's nostrils or mouth, use a clean towel to wipe the membranes off.
- If, after removing the placental membranes, you see mucus in the nose, use a suction bulb to draw it out.
- If the calf hasn't started breathing after you've cleared the mucus, tickle the calf's nostril with a clean piece of straw to get it to sneeze or cough; doing so forces the calf to take a breath.

When helping a newborn calf start breathing, position the animal so it's lying on its sternum (breastbone). Don't lift the calf up by its rear legs in an attempt to clear its lungs. The weight of the stomach and intestines will push on the diaphragm making it harder for the calf to inhale and exhale. Similarly, don't pound on the calf's chest or lay it out on its side.

After the calf is breathing normally, leave the cow and baby alone so the mother can lick the calf dry. Her rough, sandpaper-like tongue does an amazing job at removing the birthing fluids and pregnancy membranes. The mother's licking also stimulates blood circulation and gives the cow and calf time to bond with each other. If the cow was injured during delivery or is ill and doesn't show interest in her calf, use clean cloths or towels to dry the calf yourself.

Be sure the material you use to dry the calf doesn't smell of other cattle or strong bleach or detergent. Part of mother/offspring bonding depends on smell, and you don't want any powerful foreign scents to hinder the process.

Treating the navel

A newborn calf's umbilical cord can provide a gateway for disease into the body. For instance, bedding and manure can harbor organisms that can infect the calf through its umbilical cord. Newborn calves are particularly susceptible to navel infections because they have little immune protection until they consume an adequate amount of colostrum. (You can read more about colostrum and its important role in newborns in the upcoming section "Ensuring the calf gets enough colostrum.")

One of the simplest and most effective methods to help prevent infection in your young calf is to disinfect the navel.

To disinfect the navel, you need to gather the following supplies from your veterinarian or a farm supply retailer:

- **Triodine-7:** It's important to use this iodine-based antiseptic/disinfectant rather than a gentle or diluted iodine mixture, because the Triodine-7 dries more quickly and sanitizes the navel better.
- **A navel cup:** This helpful tool is a specially designed plastic handle with a cup attached to the end. It helps you disinfect properly without a huge mess. Just fill the cup with triodine-7 and then submerge the navel for a few seconds in the liquid. Don't use a spray bottle to apply the Triodine-7; achieving sufficient coverage with this method is difficult.

Treating the navel once within a few hours after birth is sufficient. Turn to "Diagnosing navel or joint ill" later in this chapter to find out what to do if the navel becomes infected.

Where's my name tag? Identifying the calves

Identifying your calves (and all your cattle) is important because it's a sign of your ownership, it makes it easier for you to reunite separated cows and calves, and it allows you (and any of your helpers) to keep track of your animals and any health work they may have received (or need to receive in the future).

Your options range from ear tags to more permanent (although not fully tamper-proof) types of identification, such as tattoos. We explain these options in the following sections.

Ear tags

The most common form of identification is a plastic ear tag. They come in several sizes and colors. You can buy blank tags and use a special tag marker to write on the tags. You can also buy preprinted tags. For promotion and ownership purposes, it's handy to have your farm name preprinted on the tags. Ear tags can be used for any age or size of animal. They're inexpensive and quick to put in (for the how-to, see Chapter 9). However, keep in mind that they can fall out or be removed.

You can insert ear tags when you treat the navel or anytime later in the life of the calf. One tag can provide a lot of information. Here are some tips for using them on calves to ease identification:

- **Use tags to identify the animal's sex.** Put tags in the left ear of male calves and the right ear of female calves. When it comes time for sex-specific treatments, you can then quickly and easily sort the calves without messy under-the-tail checks.
- **Pair up mother cows and calves by using the same numbers on their tags.** Doing so makes sure you can quickly and efficiently match them up.
- **Use matching colors to indicate the different sires.** You can then number the calves sequentially by the order in which they were born.

Light-colored ear tags with dark writing are the easiest to read from a distance, an attribute that comes in handy when you have a shy or scared calf that doesn't want you to get too close.

Tattoos

As you would expect, cattle tattooing involves sharp, needle-like projections that pierce the skin (usually in the inside of the ear) when held in place by a specially designed set of tattoo pliers. Tattoo ink that was rubbed onto the skin is forced into the puncture holes and remains visible after the wounds heal. This procedure is usually done around weaning time, after the ear has had time to grow and so is easier to tattoo.

Instead of being tattooed with words and characters, cattle receive tattooed letters that correspond to the year of the animal's birth, as shown in Table 12-1. Because I, O, and Q can be easily misread as numbers, they don't have a year designation.

Depending on your breed association rules, V and Z don't have year designations either. Before tattooing your purebred cattle, check with your specific association to confirm the ear tattoo letter codes.

Table 12-1 Ear Tattoo Letter Codes

Year of Birth	*Letter*	*Year of Birth*	*Letter*
2000	K	2011	Y
2001	L	2012	Z
2002	M	2013	A
2003	N	2014	B
2004	P	2015	C
2005	R	2016	D
2006	S	2017	E
2007	T	2018	F
2008	U	2019	G
2009	W	2020	H
2010	X	2021	J

I, O, Q, and V are not used.

Practice tattooing a piece of cardboard before you tattoo your animal. That way you can check that all the letters and numbers are oriented correctly (not backward or upside down).

Figure 12-1 shows what a legible tattoo looks like.

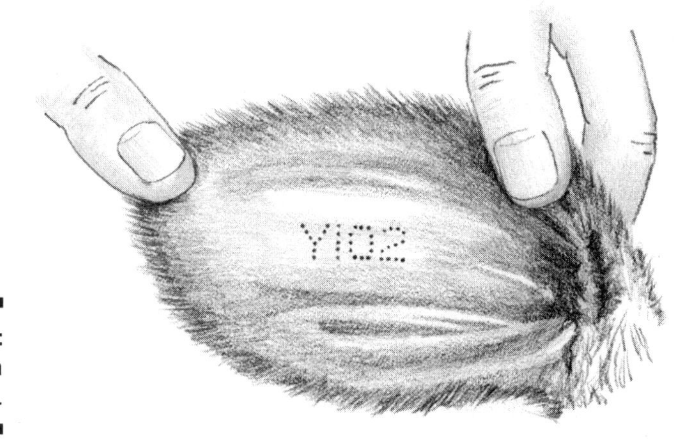

Figure 12-1: A tattoo in a calf's ear.

Ensuring the calf gets enough colostrum

Colostrum is the first milk a cow produces right after calving. It's a nutrient powerhouse with higher levels of solids, minerals, and proteins compared to the milk the cow produces later in lactation. Colostrum also contains the hormones and growth factors needed for the development of the digestive tract.

A calf is born with little or no immunity to fight off disease, so early and adequate colostrum intake is key to the calf surviving and thriving. The calf acquires the ability to defend itself against disease by ingesting the antibodies produced by its mother and transmitted through her colostrum. If the calf doesn't get enough high-quality colostrum, it will be at increased risk of illness.

By two hours after birth, a calf needs to be standing and nursing. It needs to drink 2 quarts of colostrum by six hours of age and another 2 quarts in the next six hours.

The timing of colostrum intake is so important because within six hours after birth, the digestive tract's ability to absorb antibodies is decreased by about 33 percent. At 24 hours of age, absorption is down to 11 percent of the capability at birth. At this time, the digestive enzymes also are already able to break down the antibodies.

How can you tell whether the calf is getting enough to eat when it's nursing? Pay attention to the following characteristics:

- **Physical behavior:** The best sign to look for is a calf that gives a big stretch when it gets up from the lying down position. When a calf with a content belly stretches, it arches its back, lowers its head, shakes out a hind leg, and maybe even curls its tail over the back. A hungry calf gets up and stands with a hunched, listless appearance. It may nose around on its mother, but it won't grab ahold of the teat and start nursing.
- **Feces:** The first stool of a calf is called *meconium,* and it's usually greenish black and sticky and covered in clear mucus. After this first bowel movement, if the calf is getting enough to eat, it will have frequent stools that are thick, pasty, and golden yellow in color.
- **The mother's udder:** The cow's udder indicates whether a calf has nursed. A newborn calf rarely nurses each of the four quarters of the udder evenly. If one or two quarters of the udder are smaller and somewhat flaccid compared to the other two, it's likely the calf has nursed.

Sometimes after a difficult delivery or in bad weather, a calf just won't have the strength to get up and eat. When this happens, you need to administer the colostrum by bottle feeding it to the calf. Ideally, you want to extract the colostrum from the mother cow, but you can also use extra colostrum that you've collected from other cows. You also can use a commercial powder replacement. In the following sections, we show you how to extract colostrum and bottle feed a newborn calf. And, just in case the calf won't even take a bottle, we show you how to use an esophageal feeder.

A calf that needs help nursing can be a time-consuming challenge. However, one bottle feeding may be enough to get the calf eating on its own. If not, we show you how to bottle raise a calf in the later section "Bottle raising a calf."

Extracting colostrum from the cow

If you don't have spare colostrum set aside or a commercial powder replacement available, you need to get colostrum from the calf's mom. Here's how:

1. **For your safety, restrain the cow in a head gate and chute.**
2. **If your chute side splits in the middle, only open the bottom half.**

   Opening only the bottom half allows you to reach the udder while keeping the cow somewhat contained.
3. **To get the milk out, firmly grasp the teat and pull downward with a squeezing motion.**

   It may take a few tries to get the milk flowing because the teats are usually blocked with small waxy plugs that keep infection-causing organisms out of the udder.

Bottle feeding colostrum to the calf

After you've collected 2 quarts of colostrum, try to get the calf to take a bottle. Use a ½ gallon calf bottle and have several different types of nipples at your disposal. Some calves without a good sucking reflex need a nipple with a big opening so the milk practically dribbles out on its own. Other calves do better with a soft nipple with a small hole.

Make sure your bottles and nipples are clean. Sanitize the bottles by filling them with hot water and one tablespoon of bleach and letting them soak for a few minutes. Submerge the nipples in the same solution. Rinse thoroughly before using.

After the initial bottle feeding, work to get the calf nursing from its mother. The longer the calf is on the bottle, the less it associates the udder with food. If the cow is restless or the calf is weak, you may need to put the cow in the chute so you can assist the calf safely. Slow calves may need to be pointed directly to the udder. You may even have to put the teat directly in the animal's mouth.

Feeding the calf with an esophageal feeder

If a calf is weak, it may not be able to nurse or be bottle fed. In this case, you need to use an esophageal feeder to get the colostrum or milk into the calf's stomach. Running a tube down a calf's throat may seem like a daunting task, but you can do it. You may sometimes find that "tubing" a calf is the only way to keep it alive.

An esophageal feeder or stomach tube is made of a reusable plastic pouch or bottle, which holds the fluid. This part is attached to a plastic tube with a stainless steel or plastic ball probe on the end. To insert the tube, follow these steps:

1. **Fill the pouch or bottle with colostrum or milk.**

   If the feeder has a shut-off valve, use it to stop the flow of milk until the tube is in place; if the feeder doesn't have a shut-off valve, don't attach the bottle until the tube is in position.

2. **Before inserting the tube into the calf's throat, measure the tube alongside the animal's throat so you know approximately how far in the throat the tube should go.**

   On the outside of the body, it should reach from the mouth to about where the neck joins the shoulder.

3. **With the calf standing or lying down on its sternum, hold its head steady between your legs and ease the tube into the side of the mouth and down the esophagus.**

   You'll notice two structures in the neck: the trachea with its distinct ridges and the smooth esophagus tube. You should be able to feel the ball on the end of the feeder when you move it gently up and down in the esophagus.

4. **Administer the fluids by opening the valve or attaching the bottle to the tube.**

   Let the contents move out by gravity flow.

5. **Make sure the tube has drained of all fluid before you pull it out of the calf's throat.**

   Draining the tube completely before removal keeps liquid out of the calf's trachea. Liquid in the trachea leads to aspiration, which can result in choking.

Figure 12-2 illustrates the proper positioning of an esophageal tube feeder.

Using an esophageal feeder can seem scary at first, but don't be afraid to learn. It's an invaluable tool for saving weak or ill calves.

Bottle raising a calf

If a cow is unable to produce milk or a calf is being stubborn and won't nurse, the calf needs to be raised on a bottle (see the earlier section "Bottle feeding colostrum to the calf" for information on bottle feeding basics). For the first few days, feed the calf every 8 hours; after that, you can drop down to every 12 hours.

Figure 12-2: Using an esophageal feeder.

When bottle feeding, use a high-quality milk replacer with at least 22 percent protein and 15 percent fat. For weak or chilled calves, feed warm milk replacer (but don't heat it to more than 100 degrees Fahrenheit).

A calf requires about 8 percent of its birth weight in milk or milk replacer every day. So a calf weighing 90 pounds at birth should consume a little more than 7 pounds a day.

After a few days of age, offer dry, specially formulated calf starter feed and nice grass or grass-legume hay. Make clean water available at all times after the calf starts eating solids. After the calf is eating 1½ pounds of calf starter a day, it can be weaned from milk. Bottle-fed calves can be weaned when they're around 8 to 12 weeks of age.

Dodging Calf Problems in the First Few Days

Because most post-calving deaths occur within two weeks after birth, you're well on your way if you can get the calf through the first few hours and days of life. However, you still need to watch out for several issues to help the calf get a good start in life. We explain them in the following sections.

As you monitor the health of a newborn calf, keep in mind that its normal vital signs will be in a higher range than older bovines. The normal calf temperature is between 101.4 and 104 degrees Fahrenheit. The average heart rate is 130 beats per minute, and the average respiratory rate is 56 breaths per minute. For more on bovine vital signs, see Chapter 9.

Brrrr: Dealing with hypothermia

One of the most common causes of calf mortality is *hypothermia,* an abnormally low body temperature. This condition primarily occurs during the first 48 hours following birth.

The likelihood of hypothermia escalates dramatically when any two of the following conditions are present: cold air, wind, or a wet calf. You can control these factors somewhat by selecting a mild season for calving, providing shelter and dry bedding during inclement weather, and selecting cows with good mothering abilities, including licking their calves dry quickly. But, despite your best efforts, if you raise enough calves, you'll likely have to deal with hypothermia at some point.

Understanding hypothermia

A beef calf's normal body temperature is approximately 101 degrees Fahrenheit. (See Chapter 9 for instructions on how to take your animal's temperature.) When the calf's body temperature begins to fall below 101 degrees, it begins to enter the various stages of hypothermia:

- **Body temperature between 95 and 100 degrees Fahrenheit:** At this temperature range, the calf is mildly hypothermic. It begins to shiver, and its respiration and pulse rate increase as its body tries to keep warm. When you feel the calf's hooves, they may be cold due to blood being shunted away from the extremities to preserve the core body temperature.
- **Body temperature between 94 and 86 degrees Fahrenheit:** The calf is severely hypothermic at this point. At this stage the vital organs begin to get cold and brain function can be impaired.
- **Body temperature drops below 86 degrees Fahrenheit:** The calf is near death when it reaches this body temperature; it will be nonresponsive and have limited vital signs.

If you catch hypothermia before it becomes severe, you have a good chance of saving the animal. Calves are surprisingly resilient, and results can be dramatic and rewarding. Refer to the next section to see how you can warm up a hypothermic calf.

Raising a calf's body temperature

If a calf is slightly hypothermic, all you may need to do is place it on some dry bedding away from wind and rain, and then dry it off by rigorously rubbing it with some dry towels. If you aren't sure whether the calf has nursed, it's also a good idea to provide it with colostrum by bottle or esophageal feeder (you can read about both options earlier in the chapter).

When a calf needs additional assistance to get its temperature back to normal (101 degrees Fahrenheit), any of the following methods can be effective:

- **Wrap the calf in a dry blanket.** Make sure the blanket is clean and doesn't smell too strongly of other cattle or of bleach. Strong smells can impede bonding between mother and calf.
- **Place the calf under heat lamps or under the floor-board heater of your pickup truck.** Take care the calf doesn't become overheated though. Usually when your calf gets warm enough, you won't be able to keep it under the heat source anyway!
- **Warm the calf in a heating box.** A *heating box* is a box that's big enough for the calf to lie or stand in and that can be heated to approximately 105 degrees Fahrenheit.

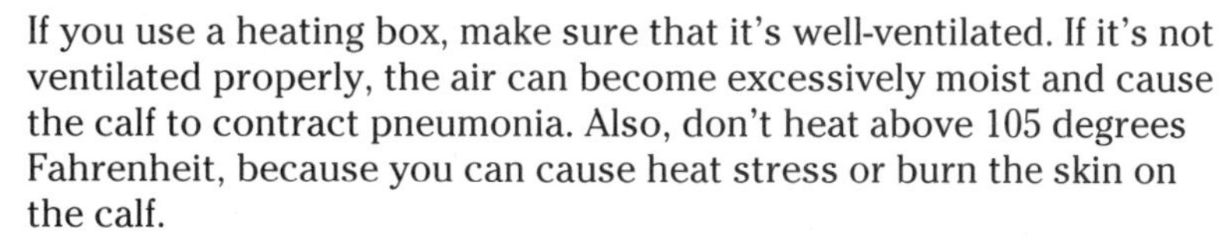
  If you use a heating box, make sure that it's well-ventilated. If it's not ventilated properly, the air can become excessively moist and cause the calf to contract pneumonia. Also, don't heat above 105 degrees Fahrenheit, because you can cause heat stress or burn the skin on the calf.

- **Place the calf in a warm-water bath.** If the calf has severe hypothermia or is nonresponsive, a warm-water bath has been shown to be one of the quickest and most effective methods of increasing the body temperature (although not the most convenient).

Diagnosing navel or joint ill

Newborn calves are susceptible to infection through their umbilical cords. Calving in a clean environment and disinfecting the navel soon after birth goes a long way toward preventing infection. (For details on prevention, refer to the section "Treating the navel" earlier in this chapter.) If a calf is infected through its umbilical cord, it may result in a disease called *navel ill.*

A calf with navel ill may

- Have a wet, swollen navel
- Be lethargic
- Avoid nursing

If the infection spreads beyond the calf's navel, it can quickly travel throughout the body and be fatal. It can also settle in the calf's joints or organs and become a chronic localized condition. After it spreads to the joints, the disease is generally referred to as *joint ill.* Calves with joint ill are lame with swollen and painful joints.

Navel ill and joint ill are easier to prevent than to treat. After the disease is present, early diagnosis and persistent treatment are crucial to a successful outcome.

If you have a calf with navel ill (or joint ill), contact your veterinarian immediately to get the appropriate antibiotic treatment for the calf. Make sure to give the antibiotic according to your veterinarian's instructions because treatment is required until the infection has been fully eliminated, which will be well after the calf seems recovered. If a navel abscess is present, the vet may also want to lance it and drain the infected tissue.

Looking out for scours

Calf *scours,* or diarrhea, is one of the most common conditions affecting young calves. Scours itself isn't a disease. It's a clinical sign of illness, which can be caused by infectious organisms, nutritional upset, or environmental factors acting alone or in combination with one another. The feces associated with scours can come in many different variations. It is watery and can range in color from yellow to green to white. It may contain clear mucus or blood.

Be diligent about preventing scours because it's one of the leading causes of death in young calves.

Deterring infection

Prevention of scours is important to the health of your herd. Here are some tips to include in your prevention program:

- **Calve outside on a clean, grass lot when weather permits.** You can keep your pasture clean either by strip grazing the pasture so the cows and calves are moved daily to a fresh section of grass or by providing a large pasture with a half acre per cow. If you use a large pasture, it's best to move the mothers and their calves to a different location away from the expectant mothers so the older calves don't spread disease to the weaker newborns.
- **If you calve inside a barn or shed, remove the old bedding after each birth.** Put down a fresh layer of lime to absorb odors, and then cover the area with clean, dry straw. Turn to Chapter 11 for details on designing a calving facility.

- **Make sure calves receive adequate amounts of colostrum.** Research has repeatedly shown that consuming colostrum reduces the likelihood of scours and other calfhood diseases. You can read about colostrum requirements in the earlier section "Ensuring the calf gets enough colostrum."
- **Provide the calf with a regular supply of milk.** If you're raising a bottle calf, feed it consistent amounts of milk on a routine schedule. For nursing calves, don't keep them separated from their mothers for long periods of time. When they're reunited after a long absence from the bottle or their mother, hungry calves may overeat. Overeating leads to digestive upset and makes the calf more susceptible to disease. The scours associated with overeating are often white, due to undigested milk passing through the gut.
- **Give the calf shelter from harsh weather.** A half-day storm that's no big deal to a cow can be mighty tough on a young calf. And hot summers can also take a toll on youngsters. They need shady places to escape to. One option to consider is housing calves in a portable calf barn during inclement weather. If the shelter has openings just big enough for a calf to fit through, it stays cleaner because no cows come inside.
- **Isolate sick calves and care for them after tending to all the other calves.** This separation reduces the chance of spreading disease.
- **Vaccinate your cows prior to calving.** Confer with your vet to determine what problematic viruses and bacteria are of greatest concern and to develop the best vaccination schedule. The specific presence of particular organisms can be determined by a combination of diagnostic lab tests, clinical signs, and necropsy results.

  Some possible viruses to immunize against include the rotavirus, coronavirus, and bovine virus diarrhea. You can also vaccinate for scours caused by *E. coli* and *Clostridium perfringens* bacteria. Head to Chapter 11 for information on vaccinating your cow herd.

Managing with supportive care

In spite of the best management practices, calves sometimes still get scours — and some may even die. Even though a disease-causing agent, nutrition, or weather may have started the scours, what most often causes death is dehydration, loss of electrolytes, and changes in pH. So although attacking the causative agent is important, supportive care is even more critical.

A calf with scours needs fluid, electrolytes, and energy. Fluids are important so body organs can properly function. And to absorb fluids from the digestive tract, the body needs the correct amount of salts, which are provided by electrolytes. Finally, to fuel the absorption of liquid and the work of the body's natural defenses, energy is required.

Observe the calf to make sure it's getting enough liquids. A dehydrated calf has sunken eyes and loses the elasticity in its skin. To check for elasticity, pinch and gently pull a 1-inch fold of skin up off the back. If the skin doesn't spring back down flat, the animal is dehydrated. Calves with lots of diarrhea and dehydration need aggressive supportive treatment.

Start supportive treatment quickly and give often. Provide the young calf 2 quarts of fluid mixed with electrolyte product two to four times a day. Your vet's office should be able to supply you with a commercially prepared electrolyte product. If you're feeding milk, delay the fluid and electrolytes until 15 to 20 minutes after feeding.

Your goal should be to keep the calf strong enough so its body has a chance to fight off the disease-causing organism. If bacteria are involved, oral or injectable antibiotics can be useful.

With salmonella-induced scours, exercise care when giving antibiotics. The antibiotics may cause the release of endotoxins from the salmonella bacteria, which may lead to the animal going into shock. Fluid therapy may be the only option in those cases.

Growing Up: Tending to the Older Calf

In what seems like the blink of an eye, the little newborn calf that could barely stand is now a 400-pound youngster dashing across the pasture. As the calves grow, their needs change. In the following sections, we provide the information you need to know about the needs of older calves, including everything from creep feeding and weaning to dehorning and castrating.

Expanding the calf cuisine: Providing creep feed

By the time a calf is 90 to 120 days old, its mother's milk is only supplying about 50 percent of the nutrients it requires for maximum growth. If you have plentiful, high-quality pastures, they're an excellent low-cost supplement to milk. However, when pastures are short in supply and nutrient value, you can supplement the calf's diet while it's still nursing by *creep feeding.*

A *creep* is an area where feed is provided for only calves; no big, hungry cows allowed. The calves get to the creep feeder or into the creep area through gates with openings big enough for them but small enough to keep cows out.

Creep feed can be grain, limit-fed protein supplements, or high-quality grazing pasture that the cows can't access. You can make your own ration or buy a premade creep feed.

Calves that are creep fed handle the transition of weaning better because they're used to eating on their own.

Developing a vaccination schedule

Vaccinating your calves is an important contributor to their overall health and value. Your main area of focus should be the viruses and bacteria associated with the respiratory and clostridial diseases. (Chapter 10 goes into detail about these disease-causing organisms.)

Work with your vet to develop a protocol that best addresses the problems your own herd faces plus any concerns that are more local or regional in nature. We provide some general guidelines in the following sections.

During the first few months

For the first few months of life, calves are protected by the antibodies they received from their mothers' colostrum. Because immunizations don't confer immediate protection and most require a booster dose, you want to vaccinate the calves while they're 2 to 3 months of age before the immunity provided by the colostrum wears off. At this time, give the calves a clostridial 7-way shot and a respiratory vaccine for infectious bovine rhinotracheitis (IBR), bovine virus diarrhea (BVD), parainfluenza-3 (PI-3) virus, and bovine respiratory syncytial virus (BRSV). (For a refresher on the proper technique for giving shots, see Chapter 9.)

Use only killed vaccines in calves nursing pregnant cows. Modified live vaccines could move from the calf to its mother and cause an abortion.

Heading into weaning

Either three to four weeks before weaning or at weaning, your calves need a booster dose of clostridial 7-way and a respiratory vaccine for IBR, BVD, PI-3, and BRSV. The clostridial 7-way at this time should also protect against *H. somnus.* In addition, immunize your calves for *Pasteurella.* If you're going to give the booster shots before weaning, do so at least 3 weeks prior to weaning. This timing allows the body to mount an immune response to the vaccine.

Evaluate the stress caused to your cattle when they're handled to decide whether it's better to work them three times (two rounds of shots plus weaning) or just twice (two rounds of shots with the second dose at weaning).

Treating future moms

Heifers that you'll keep as mother cows need to be vaccinated against brucellosis between 4 to 12 months of age. This immunization must be given by a licensed vet, and that vet must also give the vaccinated heifers an official ear tag and tattoo.

Handling horns

The best way to deal with horns is to not have to deal with them; in other words, consider breeding for *polled* (naturally hornless) cattle. For more about the inheritance of the horn trait and how you can breed for polled cattle, see Chapter 3.

If your calves do have horns, you want to remove them — unless, of course, you're raising Texas Longhorns or another breed associated with fancy horns. In general, cattle with horns are worth less money at sale time and are more dangerous to humans and other animals.

To reduce the risk of complications, dehorn calves at a young age. You also want to dehorn in the early spring or later fall to avoid the fly season and reduce the risk of infection due to flies. The calf needs to be properly restrained to prevent injury to you and the animal. For smaller calves that are less than a month old, you can simply lay the calf on its side and have a helper hold it down. Older calves need to be placed in a head gate and chute. (Turn to Chapter 4 for details on this type of equipment.)

Proper dehorning removes the horn or the *horn bud* (the name of the small developing horn before it's attached to the skull) and prevents future regrowth. Methods to dehorn include the following:

- **Chemical dehorning:** This method is appropriate for calves a day old up to three weeks of age. While wearing gloves, apply petroleum jelly around the horn bud. Then dab the chemical caustic material (commercially available from your vet or farm store) inside that area. The chemical destroys the horn-producing cells, causing the bud scab to fall off within a week or two. The pain associated with this dehorning technique is minimal.

  Don't let the calf around its mother until the chemical paste dries. The paste could get on her udder when the calf nurses, causing burning and irritation of the udder. Also, keep the calf out of the rain for several days so the chemical doesn't come in contact with the eyes or skin and cause burning.
- **Tube dehorning:** This removal technique is an option for calves that are younger than two months and have horn buds shorter than 1½ inches. To perform this dehorning, you use a sharp dehorn tube slightly larger than the horn bud. You place the tube over the bud and then push through the skin. After penetrating the skin, lean the tube to the side and scoop out the bud.
- **Hot iron dehorning:** This technique can be used on horns shorter than 1 inch. Hold a hot iron to the horn and surrounding skin. After 20 seconds of contact, the skin should be burned white and will no longer produce horn cells. The current horn should die and fall off within a month or

so after treatment. Take care not to leave the hot iron in place too long because excess heat can damage the brain.

- **Barnes dehorning:** This technique can be used on calves from 2 months up to 1 year of age. To remove a horn with this technique, you use a Barnes dehorner to cut through the base of the horn and the surrounding skin. It takes quite a bit of strength to completely close the dehorner through the skin and horn base.

Before performing the tube, hot iron, and Barnes dehorning methods, ask your vet to show you how to give your cattle local anesthetic, like lidocaine, to block pain during and immediately after the dehorning. And be prepared for lots of blood with these procedures. You may need to cauterize the arteries to stop the bleeding. Surgery sites should be treated with a wound spray and a fly repellant that's safe for use on open sores. Also, whatever method you use, be sure all your equipment is disinfected and in good working order. Chlorohexadine is a good choice for disinfecting.

No matter how skillful or careful you are, dehorning is still stressful for the animal. You can eliminate the need for this procedure by using homozygous polled bulls as herd sires.

Castrating young bulls

Castration is the removal of a bull's testicles either by surgical or nonsurgical methods. This practice is done to improve the animal's temperament and make it safer for humans and other cattle to be around. Steers (castrated males) almost always bring a higher price at market than bulls. We explain methods for castrating in the following sections.

Banding

Banding, or elastration, is used to castrate calves that are 1 month old or younger. With this method, you place a strong, thick rubber band around the top of the scrotum where it connects with the body. You expand the band and slip it over the scrotum with a hand-held elastrator tool. As you put the band on the calf, be sure the testicles are pulled down in the scrotum so the rubber band is between them and the area where the scrotum connects to the belly. As you release your grip on the elastrator, the band should tighten, and you should be able to slide the tool away from the scrotum while the band stays in place. The blood flow to the scrotum and testicles is stopped so they atrophy away in a few weeks. This is a simple technique you can do yourself.

Clamping

With the *clamping* method, you use an *emasculator* (a clamping tool) to crush the spermatic cord. By crushing this cord, you prevent the testicles from getting blood flow and signals from the nervous system, so they atrophy and become nonfunctional. You can clamp your bulls once they are at least a few months old and up until they reach about 800 hundred pounds. You can do this procedure yourself if you have secure handling facilities, or you can hire your vet to do it. To clamp your young bull yourself, follow these steps:

1. **Securely restrain your bull in a working chute and head gate.**
2. **Have an assistant hold the tail over the animal's back to prevent it from kicking.** Holding the tail over the back alters the animal's balance so that it can't raise its leg to kick.
3. **Push the testicle down in the scrotum and clamp the tool closed about 2 inches above the testicle, directly over the spermatic cord.**

   The spermatic cord goes from the top of the testicles through the narrow part of the scrotum and into the body. Leave the tool in place for at least 30 seconds, release, and then repeat the procedure one more time to ensure the cord is completely clamped.
4. **Complete the second side using the directions in Step 3.**

Surgical castration

Surgical castration involves using a knife to make an opening in the scrotum through which the testicles are removed. Another surgical method involves cutting the cord as close to the body as possible and then pulling the testicles from the body.

Surgical castration should be done by a vet and is often reserved for larger (over 800 pounds) bulls in which banding and clamping are no longer viable options.

Easing the stress of weaning

Weaning is the process of separating a calf from its mother. This process usually takes place when the calves are anywhere from 5 to 8 months old. The decision of when to wean depends on your marketing plans and feed resources.

If, for example, you're selling your calves as weaned feeders, try and time their weaning with the highest demand and sale prices. If you're retaining ownership, however, you have more flexibility in determining your weaning time. For instance, when feed for your cows is plentiful but high-quality calf feed is harder to come by, it may make sense to let the calves nurse awhile longer. The cow can convert poor-quality feed into higher-quality milk for her calf. Just take care not to let your cow's body condition become poor (for more on body condition, see Chapter 11). If you're under drought conditions, you can wean your calves early and feed them harvested feed in an attempt to conserve your pastures.

Weaning time is stressful because the calves not only have to adjust to a diet without milk, but they also have to adjust to a new social structure without the leadership of the adult cattle. The key to successful weaning is smoothing the transition so the cattle keep eating and are thus less susceptible to disease.

You can wean your calves using fence-line weaning or separation weaning. However, studies comparing the two methods have found that fence-line weaned calves gain more weight in the time after weaning and have less respiratory disease than separated calves. We explain both techniques in the following sections.

Fence-line weaning

With fence-line weaning, the cows and calves are physically separated so the calves can't nurse but the pair can still see, touch, and hear one another. Fence-line weaning is growing in popularity thanks to its ability to ease the stress of weaning.

If you have enough space — you need two adjacent pastures and an adjoining handling facility — one nice way to do the fence-line weaning is to move the whole herd to a fresh, highly palatable pasture. Follow these steps:

1. **Leave the cows and calves in a new pasture for a few days.**

   Leaving the cattle on the pasture for a few days allows the calves to become accustomed to their new surroundings. The calves will stay on this pasture in the end, but the cows will move to another new pasture next door.

2. **Move the whole herd to the handling facility.**

3. **After the animals have settled in and the calves have had one last chance to nurse, open the gate to the cow's new pasture. Move the cows to the pasture, leaving the calves behind in the handling facility.**

   It's much easier to get cows to leave one location and move to a new one than calves. This method allows you to take advantage of calves' natural tendency to hang back in the handling facility and let the cows out first.

4. **When all the cows are gone from the handling facility and are in their new pasture, open the gate so the calves can return to their original pasture.**

   You can expect some pacing and bawling for the first few days, but because the mothers and calves can still be in contact with one another, they'll spend a lot more time eating and much less effort pacing than if they were totally separated.

Be realistic about your fence when fence-line weaning. It doesn't have to be strong enough to corral a charging elephant, but four rusty barbed-wires won't be enough either. The cow-calf bond is often stronger than a run-down fence. Unless, you have a wily calf or cow with a bad attitude, a good woven wire or a seven-strand electrified high tensile fence should suffice. Chapter 4 has more information on fencing.

Separation weaning

If you don't have the space or facilities for fence-line weaning, you have to do a total separation weaning where the cows and calves have no contact with one another. Of course, with this method, you still want to make it an easy process for your cattle. Some things you can do include the following:

- **Remove the cows from the calves, not vice versa.** When calves can stay in a familiar environment, they have one less adjustment to make.
- **Keep calves on the same diet during weaning.** If you need to introduce new feeds, try to give calves access to them in a creep area before weaning. Check out the earlier section "Expanding the calf cuisine: Providing creep feed" for details.
- **Take advantage of the calves' tendency to walk the fence perimeter.** Capitalize on this behavior by putting feed and water in bunks and troughs perpendicular to the fence so the calves easily encounter their feed and water source.
- **Attract calves to their water source by letting them hear water running into the trough.** This enticement is especially important for calves used to drinking from streams or ponds.
- **Maintain strong fences.** Calves and cows weaned by separation are more likely than fence-line weaned animals to test the fence.
- **Move the cows out of hearing range of the calves (if possible).** The animals will be less stressed if they can't hear one another vocalizing.
- **Schedule management practices like vaccinating, dehorning, castrating, and branding at least three weeks before weaning.** Doing so gives the calves time to recover before the stress of weaning.
- **Try to wean your calves during a period of mild, dry weather.** As with all aspects of raising cattle, you get the best results when you work with nature and not against it.

Part IV
Realizing Your Cattle Business Potential

"I guess we should have expected this when he named the cow, 'Stroganoff.'"

In this part . . .

This part focuses on the business end of raising beef cattle. Chapter 13 provides an overview of how you can turn your green pastures into green dollars by raising stocker calves, caring for other people's cattle, or specializing in 100 percent grass-fed beef. Chapter 14 delves into the fascinating world of cattle shows and touches on other ways you can promote your cattle business. Chapter 15 highlights the details of managing a beef business, including a review of the many regulations you need to follow.

Chapter 13

Turning Your Extra Pastures into a Money-Making Business

In This Chapter

- Raising stocker calves
- Caring for someone else's cows
- Producing 100 percent grass-fed beef

Do you have acres of fields that you're tired of mowing? Do you want to do something productive with your land, but you've found that it's too rough for traditional row crop farming? Well, you're in luck. You have several options. You can use your pasture to feed growing calves, to look after another farmer's cows, or to produce 100 percent grass-fed beef to sell or butcher for your own table.

In this chapter, we go over the benefits, drawbacks, and requirements of the different options so you're more likely to have an enjoyable and profitable experience.

Growing Stocker Calves

Stocker calves are a great choice if you have extra grass or if you want to raise cattle but aren't ready for the commitment of caring for cattle all year. *Stocker calves* are cattle of either sex that weigh between 400 and 500 pounds and are fed a high-forage, pasture diet until they reach 700 to 900 pounds. At that point most stockers go into a feedlot, where they eat a higher-concentrate finishing diet until reaching butcher weight of 1,000 to 1,400 pounds. They're often recently weaned calves, but you can also raise yearling cattle as stockers; just take care to avoid buying stunted animals that don't have much growth ability.

The goals of feeding stocker calves a high-forage diet are to have them grow in size without becoming too fat and to achieve this weight gain in an economical manner.

In the following sections, we explain everything you need to know to get started raising your own stocker herd, including choosing a season, picking the best cattle and source of cattle, planning for a buyer, and preparing your farm and animals.

Deciding on a season

You can raise stocker calves any time of year as long as you have enough forage available. The ideal time frame depends on which part of the country you live in and the availability of animals suited to being raised on pasture.

Spring and summer grazing

Purchasing calves in the spring and grazing them until fall or when your grass runs out — whichever comes first — enables you to take advantage of the bountiful pastures found in many parts of the country during May, June, and July.

The drawback of grazing your stockers during this season is that many of your neighbors may have the same cattle-rearing plans. As a result, finding healthy, reasonably priced calves may be difficult. You'll find this scenario to be especially true if most of the calves in your area are born in the spring, because then the calves in your geography at this time are either young, nursing calves or yearlings, which are often more suited to a concentrate-based finishing diet (see Chapter 5 for details on finishing diets). You may need to line up your source of stocker calves well in advance to avoid being faced with a limited selection.

Winter grazing

Southern states with mild weather have a better supply of forage through the winter months than their northern counterparts. Calves weaned in the fall are a good fit for these programs. If you live in an area where winter wheat is a prevalent crop, you can purchase calves in the fall to graze the wheat leaves and then take the calves to the feedlot for a higher-concentrate finishing diet in the spring.

Winter grazing can be profitable because you can often find a good supply (and usually a corresponding drop in price) of newly weaned, spring-born calves. It's also less expensive to graze over the winter than it is to feed harvested forages. Be sure you have a good supply of pasture because buying hay gets expensive.

Identifying a source

Before heading out to select your stocker calves, you must determine where you can buy. Luckily, you have several options for sourcing your stocker calves. You can do any of the following:

- **Raise your own.** One of the best things about using your own calves as stockers is that you can control their diet and health program prior to entering the stocker phase. This control helps reduce the likelihood of stress and illness while increasing the potential for growth. However, keep in mind that the pastures for stockers need to be higher in quality than what cows need to meet the protein and energy needs of the growing stocker animal.
- **Buy direct from cow/calf producers.** This option allows you to discover more about the background of the calves as opposed to buying at auction. However, you may have a difficult time finding calves direct from the farm that are available at the time and size you need.
- **Purchase them at an auction.** Going to an auction allows you to select from a large number of cattle at one time. However, keep in mind that cattle from these venues are exposed to lots of germs and stress that can lead to illness.
- **Raise them on contract.** If you don't have the capital to purchase stocker cattle, you can agree to raise them for someone else and get paid for it. Be sure you have a signed, written agreement detailing your responsibilities, the obligations of the owner, and the payment schedule. Your involvement can be as simple as providing the grass to being totally responsible for all phases of feeding, healthcare, and marketing. You can arrange to be paid a lump sum, on a per-head basis, or for every pound of gain. For more tips on creating a contract for this purpose, check out the later section "Tending Cows for Absentee Owners."

Selecting good stocker calves

If you opt to grow stocker calves on your farm, you have to consider several different factors when purchasing the calves, including their selling and starting weights and their profitability. We discuss each in the following sections.

Paying attention to weight

When purchasing stocker calves, start with the end in mind. In other words, ask yourself what kinds of calves buyers want and don't want. If, for example, your region shows no market for calves over 750 pounds (which is often the cut-off weight for calves to still be considered feeders), starting with a 500-pound calf that may gain 2.5 pounds per day over five months will produce an 875-pound animal that no one really wants to buy.

Because the cattle marketplace varies greatly from region to region and even throughout the year, talk with your local extension agent and read sale reports on the Internet or in agricultural papers to become informed about the desired weights and types of calves in your area. (To find contact information for your local extension agent, go to `www.csrees.usda.gov/Extension/USA-text.html`.)

Weather, feedstuffs, and genetics all influence a calf's average daily gain (ADG). For your growing animal's health, it should gain at least 1.5 pounds daily. The upper end of weight gain you can usually expect on a forage-based diet is 3 pounds. Keep records of your cattle's starting and ending weights plus their number of days on feed so you can determine the ADGs your cattle are achieving. For example, a stocker that you start on pasture in May weighing 500 pounds that grazes through August and reaches 700 pounds has an ADG of 1.62 pounds (200 ÷ 123 = 1.62).

Of course you want to buy healthy calves (Chapter 7 gives you the scoop on purchasing healthy cattle), but it won't hurt if they're a little "green." By *green* we mean that although they're healthy, the calves are lean and lanky as opposed to plump and round. Fat calves have probably had access to high-energy concentrates, and their performance will likely suffer if switched to a lower-energy forage diet.

Making sure your calves are profitable

A good stocker calf needs to be profitable. The biggest contributor to profitability that you can control is the purchase price. If the cost of the calf and the cost of its feed is more than the sale price you receive, you'll lose money (and this loss doesn't even take into account the cost of equipment, transportation, medication, taxes, insurance, and labor).

For example, a 400-pound calf purchased at $1.30 per pound costs $520 dollars. If you feed the calf for 150 days, and it gains 2.25 pounds per day, it will gain a total of 337½ pounds. The price of gain can vary considerably, but for this example we assume it costs $50 for every 100 pounds of gain. So it costs $168.75 ([$50 × 337.50] ÷ 100) to grow a calf from 400 to 737.5 pounds. Your purchase and feed costs are $688.75, so you would need to sell your stocker for at least $0.93 per pound to cover the cost of the calf and feed ($688.50 ÷ 737.5 = 0.93).

Be conservative when estimating your selling price so you don't overextend yourself when buying.

Keep track of all your costs so you can get an accurate accounting of expenses beyond just the two big ones (cattle and feed cost). You also need to be financially and mentally prepared for the possibility of losing calves to illness or death.

Planning for a buyer

Do yourself a favor and have a buyer or marketing option in place before you even start raising your stocker calves. You have a better shot at making a profit if you know the buyer and the price you'll receive for your cattle when you're ready to sell them. Some of the options for selling your stockers are as follows:

- **Consign them to an auction specifically for health- and source-verified cattle.** Buyers are willing to pay more for cattle that are sold with some sort of guarantee regarding how they were raised and where they came from. Usually the seller provides a signed document detailing the vaccination and feeding program the calves were raised with.
- **Sell them directly to small feedlots in your area.** Feedlot owners are like most auction buyers; they're willing to pay more for calves that are health and source-verified because it means less work and potential sickness for them to deal with.
- **Work with other local producers to pool your cattle into a larger group and sell them through an Internet auction.** Many local sale barns don't have the space in their sale arena to group more than 15 head of cattle at one time. However, many buyers from big feedlots want to buy a semi-truck load of cattle (50,000 pounds) at one time and are willing pay a premium for groups of cattle that size. When you sell through the Internet you can group your cattle in bigger lots and reap higher prices.

Preparing your fields, facilities, and cattle

Before you can populate your pastures with young, healthy cattle to raise as stockers, you need to do a little homework to make sure your pastures and facilities are ready for them. Additionally, you need to make sure the cattle you've selected are healthy and have received any vaccinations or other preventive health treatments they may need. We show you how best to prepare in the following sections.

Deciding what to plant

Perennial pastures of legume and grass mixes are excellent sources of nutrients for stockers from May through early July, especially in the northern two-thirds of the country. Herdsmen living in those areas can use annual ryegrass for early spring and fall grazing. For the deep South, warm-season perennials like Bermudagrass and bahiagrass are options for winter grazing. Check out Chapter 6 for specifics on determining how much pasture you need.

If you plan to grow wheat for grazing, sow it in early September in most areas. You can have the cattle start grazing it four to six weeks after planting as long as you can see 6–12 inches of plant growth.

To reduce the negative impact of grazing on the yield of the wheat grain, remove the cattle from the pasture before the wheat plant's first hollow stem is 1.5 centimeters in length. You'll see this length sometime in the spring, depending on the weather, grazing pressure, and variety of wheat in use.

To determine the presence of the hollow stem, dig up five plants from an ungrazed area of the field (grazing delays the first hollow stem). Select the largest sprouts on the plant, and then split their stems. If the length of hollow stem is greater than 1.5 centimeters (about the diameter of a dime) from the roots to the developing wheat grain kernel, the wheat is at first hollow stem. Continued grazing of these plants decreases wheat grain yields.

Setting up your facilities

Facilities for stocker cattle are simple. Because most of the grazing takes place during warm weather or in temperate climates, cattle can stay out on pasture without barns. However, you do need the following items:

- **A simple corral and handling facility** for loading and unloading cattle. Chapter 4 provides more information on facilities.
- **A good perimeter fence around the pastures** for the safety of the cattle and your peace of mind. See Chapter 4 for fencing ideas.
- **An ample and dependable water source** is a must for any cattle operation. Head to Chapter 5 to read about the importance of water in the cattle diet.
- **Troughs or feeders** if you think you may end up giving your cattle supplemental feeds. You can read about supplementing in Chapter 5.
- **Shade sources** to provide relief from summer heat. Chapter 4 shows how to create shade for your herd.

Taking care of your cattle

Unless you're well acquainted with the source of your stocker calves and can verify they've received all necessary healthcare treatments, be prepared to give them the needed vaccinations and parasite control. Raising stockers is only profitable if your calves stay healthy and grow, so don't neglect these important steps. For guidelines on preventive healthcare, check out Chapter 9.

Tending Cows for Absentee Owners

Cows can be expensive to buy. If you want to raise cows but don't have the money to get started, investigate the possibility of *custom grazing,* which is the practice of tending cows for absentee owners on your pastures.

Custom grazing can give you hands-on experience in caring for livestock and pastures while getting paid to learn! Or, if you're in the enviable situation of having more feed than your cows can consume, you can work out a deal to manage another producer's herd on the extra fields. You can manage these herds in exchange for cash or for some other sort of compensation, such as trucking services.

If you have abundant feedstuffs, don't let them go to waste. Finding hungry cattle to fill your pastures may take some effort, but you'll almost always find some out there just waiting for you.

Because mature, dry (nonlactating) cows in their second trimester of pregnancy have the lowest nutritional needs of any type of cattle, they're usually the easiest to custom graze. So we recommend you start with them.

If you're ready to find some cows to custom graze, check out the following sections, which provide details on selecting the appropriate number of cows based on your resources, finding the cows you want, and creating a contract.

Matching the number of cows on your wish list to your resources

How much acreage do you need per cow when custom grazing? It depends on the following:

- **Pasture yield:** This factor is influenced by the weather, soil amendments, and grazing management.
- **Dry-matter intake:** This consideration is impacted by cattle size.
- **Efficiency of pasture use:** If you rotational graze, you can get more production per acre than if you give the cows total access to the entire pasture at all times.

Turn to Chapter 6 for more information on matching your pasture resources to your cow numbers.

Keep good records of grazing dates, cattle numbers, and precipitation amounts so you know the specific capability of your pastures.

Until you've had the opportunity to gather data for your own farm, here are some general requirements for a 1,000-pound dry cow:

- **Bluegrass and white clover pasture:** In May, one cow needs about 0.6 of an acre, but during the heat of summer when these forages only grow a small amount, she needs around 5.5 acres. In the fall, she needs 2.5–3.5 acres.

- **Orchardgrass pasture (with nitrogen fertilizer):** For grazing in May, a cow needs 0.8 acres; in July she needs 2.8 acres; in September she needs 1.8 acres; and in October she needs 9.2 acres.
- **Summer-seeded turnip fields:** During the late fall and early winter, one cow needs 0.5–1 acre per month.
- **Stockpiled fescue pasture:** In October through December, you can figure that each cow needs about 1 acre each month.
- **Cornstalks:** You can start allowing your cows to graze cornstalks immediately after harvest in the fall. One cow needs about 0.6 acre per month. The nutrient value of cornstalks can decline quickly, so the cows may need to be supplemented with good-quality hay.

If you have a lot of corn grain left in the field, limit your cow's access until she adjusts to having grain in her diet. Cows really like corn and will overeat and could founder (for more about founder, see Chapter 10).

Finding cattle (and their owners)

Matching your cattle-raising dreams with another person's financial backing can be a win-win situation. Here are some of the scenarios that may precipitate an owner's need to custom graze his cows:

- **Lack of time or space:** Many people want to have a few cows but, due to other commitments or lack of suitable space, can't be responsible for the animals' daily care.
- **Short on feed:** Some cattle producers are short on feed due to drought, cold weather, or herd expansion, so seek out these folks.

Don't wait for opportunities to come to you; go out and find them. Inform your neighbors, vet, and extension agent of your interest so they can spread the word or provide you with any leads. Also, put ads in agricultural publications saying that you're looking for cows to graze. Use the Internet to reach out to potential cattle owners. Place ads under the heading "Pasture for Lease," or take advantage of cattle organization websites that match up drought-stricken producers with farmers who have extra pasture.

Creating a contract for services

When contracting with a cattle owner to custom graze your pastures, write up a dated and notarized contract detailing the terms of the custom-grazing agreement. Such an agreement protects your interests by clearly outlining all parties' responsibilities and financial obligations. The contract should include the following:

- **Names and addresses** for both you (the livestock feeder) and the livestock owner. Include anyone else with a financial interest in the cows (such as a bank).
- **Date of cow delivery and pickup** with documentation of each cow's identification (brand, ear tag, tattoo, and so on).
- **Agreement on the body condition score (BCS) or weight** of the cows at arrival and departure. Including this information helps ensure you receive healthy cows and shows you cared for them properly on your end. See Chapter 11 for details on BCS.
- **Requirement that calves have been weaned** at least a week before cow delivery. You don't want a bunch of freshly weaned cows mashing down your fences looking for their calves.
- **Financial responsibilities statement** that spells out the pay rate per day, payment schedule, and provisions for extra payment in the event that you have to care for any unexpected baby calves. Also cover who will pay for any vet fees, medications, insurance, taxes, and transportation costs. Speaking from personal experience regarding the payment schedule: Don't let the cows leave until the final payment is made.
- **Description of the details,** including feed ration, feed amount, pasture space, and shelter you will provide.
- **Agreement regarding death loss.** If animals can't be accounted for, how will the owner be reimbursed? Will a vet examine animals to determine a cause of death? Which party will be responsible for carcass disposal fees? At what point will the owner cease to be responsible for death loss, causing the financial responsibility to fall onto the feeder?

Raising 100 Percent Grass-Fed Beef

You can take advantage of the growing interest in grass-fed beef by raising your own grass-fed cattle. Some consumers are interested in 100 percent grass-fed beef because they believe it has nutritional benefits not found in cattle that have eaten grain. Other consumers feel that cattle raised entirely on forage are eating a diet that's more natural and sustainable than cattle that eat grain. For whatever their reasons, more consumers are seeking out grass-fed beef and, at times, the demand outstrips the supply.

No official definition or government regulation yet dictates what constitutes grass-fed beef. As a result, you need to decide how you can best raise your cattle in a healthy and natural manner and then be clear with your customers about your feeding and management practices.

Considering taste preferences

Not everyone likes the flavor of grass-fed beef. We have customers who are devoted to our 100 percent grass-fed beef, but others have tried it and decided they prefer the beef raised on a grass-and-grain diet. Grass-fed beef is almost always leaner than beef from grain-fed cattle because growing and feeding high-energy forage is economically and agronomically more difficult compared to feeding grain. The plentiful energy in grain allows the cattle to deposit more fat within their muscles (*marbling*) and around their muscles. Properly raised grass-fed beef can be just as tender as grain-fed beef, but because it has less fat, the cook needs to be careful not to overcook the meat or it will dry out. When comparing the taste of the two types of beef, the 100 percent grass-fed has a bit more of an intense beef flavor than the grain-fed.

For instance, a steer or heifer that lives in a dirt feedlot and is fed a strictly forage diet could be described as grass-fed. Being grass-fed doesn't mean the animal can only eat grass; other plants like legumes and turnips can help balance out the diet. Harvested forages are okay, too. On our farm in Indiana, we have a good forage chain, but the pasture pickings get slim in February and March, so we supplement with hay.

Raising quality grass-fed beef isn't a simple matter of putting some calves out to pasture and having them grow some. To consistently produce tasty, tender, high-quality grass-fed beef takes a lot of effort.

In the following sections, we explain how to locate customers for your grass-fed beef, what to feed your grass-fed beef, and what types of cattle are best for your goals.

Locating customers

Your first consumer of grass-fed beef should be yourself. Before you try to find customers, make sure you're producing an acceptable product. It's a tough job, but someone needs to do quality control on those burgers and ribeyes! The grinder takes care of any potential toughness on ground beef, but steaks and roasts can be hard to chew if you don't feed and manage your cattle properly. After you feel confident about the quality of your beef, you can start marketing it. Chapter 15 is the place for all the details regarding merchandising ideas and rules.

Even though demand for grass-fed beef is growing, it's still not a mainstream food item. Your best bet for reaching customers is to have a website marketing your beef; that way, when people use the Internet to search for sources of grass-fed beef, your site will appear on the results page. If you have a Community Supported Agriculture (CSA) food program in your area, see whether the

CSA members are interested in having your beef available in addition to their weekly produce allotment. Health-food stores can be another marketing outlet for grass-fed beef.

The website www.eatwild.com is a popular site for farmers advertising their grass-fed meats. During your merchandising efforts and selling opportunities, be sure to promote the fact your beef is 100 percent grass-fed.

Supplying proper pasture feedstuffs

One of the hardest parts of raising grass-fed beef is keeping your cattle on an ever-improving plane of nutrition. When cattle are eating forages (that are often lower in energy than grain), it's especially important that every day's diet is at least as good as or better than the previous meal. Feeding grain is an easy and dependable way to give your cattle higher and higher amounts of energy. When it comes to an all-forage diet, however, the changing seasons plus the unpredictable weather can make it challenging to consistently grow enough high-quality forage for raising beef on a 100 percent forage diet.

The grass-fed diet varies depending on each individual's resources and climate. But here are some suggestions:

- **Winter:** If your calves are weaned in the fall, you can overwinter them on annual winter grass pastures like rye or wheat.
- **Spring:** In the spring, calves gain more than enough weight on perennial pastures of a legume and grass mix.
- **Summer:** During the hot summer months, cattle can graze summer annuals like sorghum-Sudan or grazing corn (grazing corn is a type of corn that doesn't develop corn grains on its ears or grazing takes place before ear formation).
- **Fall:** In the fall, cattle can graze turnips or high-performing cool season perennial pastures, such as alfalfa and orchardgrass mix.

For emergency situations like drought or crop failure, have some high-quality alfalfa hay in your barn. It's palatable and nutritious but, unfortunately, expensive and hard to find in some parts of the country.

Determining what type of cattle to raise on an all-grass diet

Some producers get quite opinionated about which breed of cattle is best for grass-fed beef. In fact, we have found just as much variance within a breed as we have between breeds. Some cattle are just better suited for finishing on a forage-only diet than other cattle. (*Finishing* means feeding an animal so it deposits fat in and around its muscle. A properly finished animal is more likely to yield meat that's tender, juicy, and tasty.)

You want to select small- to moderate-framed, early-maturing, easy-fattening cattle because they require less feed for maintenance and growth and are easier to finish on grass. Grass-fed heifers typically weigh 900 to 1,000 pounds at butchering; steers are about 100 pounds larger. Larger cattle, such as those that weigh in at closer to 1,400 pounds when finished, do much better on a higher-energy concentrate ration because their larger size takes more energy to maintain and grow.

Because heifers tend to put fat on more easily and quickly than steers, they're often the better animal to raise on all-grass diets.

A good approach is to look at each animal when deciding the most appropriate feeding method. For instance, the cattle that have a plump, fleshy appearance can be slated for a grass diet, and the lanky, lean animals can have pasture plus grain. It's wise to use breeds that are recognized for marbling and fattening ability, such as the Angus, Hereford, and Shorthorn. Then you can use Expected Progeny Differences (EPDs) to select the best performers in terms of carcass merit within the breed. Flip to Chapter 7 to learn more about EPDs.

Your cattle may be ready to butcher after being on a high-energy finishing ration for at least 60 days. During this time, the animals should gain at least 2.5 pounds per day. At this level of gain, the cattle won't be overly fat but should show signs of fat deposition in the brisket and around the tailhead. If your animal's brisket is nothing but skin without any thickness or mass, it needs more time to finish. Also you should see pads of fat on either side of the tail where it connects with the body. The animal should be nicely filled out with easy-to-see muscle definition. If the animal is skinny or bony, it isn't ready for butchering. Depending on your feeding program, grass-fed beef can be anywhere from 16 to more than 24 months in age before they're ready to be butchered.

Chapter 14
Showing and Selling Cattle

In This Chapter

- Getting ready to attend a show
- Presenting your animal in the best light
- Investigating selling options for your cattle

Across the country, kids and their families spend thousands of hours working with show cattle and taking them to exhibitions. Livestock projects provide opportunities for children to learn valuable lessons of hard work, patience, and persistence. And because shows take place from coast to coast, they give you a chance to meet new people and see the country while showing your cattle. In fact, shows and fairs are an excellent way to promote your breeding stock and develop a reputation for your farm, which can lead to on-farm sales or better prices at consignment auctions.

If you aren't interested in getting your cattle all gussied up for exhibition but still want to find ways to promote your animals, performance tests and sales may be better options for you. Marketing your cattle through these means takes time and work. And you should make the effort to do it right because doing so can translate into higher prices and repeat customers for your cattle business.

This chapter takes a look at the steps involved in getting ready for a cattle show. We also discuss what you need to do to make the most of your time at the exhibition. We wrap up the chapter with a look at some other ways of selling your cattle.

Preparing Yourself and Elsie for an Exhibition

The time you actually spend in the show ring with your animal under the watchful eye of a judge is fleeting compared to the many hours of preparation leading up to those moments. To make the most of your opportunity in the show ring, spend some time deciding what you want to accomplish by showing your cattle and which shows best mesh with your goals. Also, do everything possible ahead of time to give your cattle the chance to perform their best.

In the following sections, we provide information on figuring out your goals, discovering shows to enter, training your animals to put on a good display, and packing your supplies for the trip.

Be sure to enjoy the journey along the way, and look at the time you spend practicing with your animal as an opportunity to become better at feeding and training it for optimum health and performance.

Setting your goals

Because showing cattle is expensive and time-consuming, have a well-thought-out purpose in mind before you start the process. With a goal in mind, all the months of hard work preparing for the show will have meaning beyond the few moments you spend in the show ring.

Some of the reasons you may have for showing your cattle include the following:

- **Promoting your breeding stock:** If you're raising cattle to sell to other breeders or for youth projects, shows and exhibitions are an excellent method of building awareness of your cattle. Advertising class winners and champions in print and online after the show helps further build credibility for your program.
- **Educating yourself or your family:** Preparing cattle for show offers hands-on training in livestock nutrition and handling. It's a great opportunity to learn how to plan a project, manage a budget, coordinate paperwork (such as health certificates and show entries), visit new places, and meet different people. You also have the chance to perform under pressure while working to show off your animal to its best advantage. Many youth shows offer a broad range of educational competitions in addition to just showing cattle.
- **Having fun:** Some of our most lasting memories from growing up are of the times spent working with our show animals and of the good times we had at the fairs. Showing cattle is an activity the whole family can participate in and enjoy.
- **Making money:** We put this reason to show cattle last because it's probably the most unrealistic goal. Although shows usually award some prize money, the amounts provided usually aren't enough to offset the entry fees, the money for transportation to the show, the expense of the equipment and feed, and the cost of the cattle themselves.

Fairs for participants in youth programs like 4-H or FFA (Future Farmers of America) often have an auction in conjunction with the show. Businesses and individuals support the kids by purchasing their animals for more than their market value. Except for a few rare cases at major shows, most kids net only a modest amount after all the expenses are paid.

Whether you're showing an animal you raised or looking to exhibit a purchased animal, Chapter 7 gives you some tips on the characteristics of a good show calf.

Finding the right show for you

To achieve your goals for showing cattle and have more fun along the way, attend the types of shows that are best for your purpose. After you start looking, you'll be amazed at the number of different types of exhibitions and opportunities available for you and your cattle.

Cattle shows are divided into two general categories:

- **Open shows:** These shows are open to all exhibitors, regardless of age. You can show your own cattle or hire someone else to do it for you.
- **Junior shows:** These shows have age and ownership restrictions. Participants can be no older than 21 years of age (though sometimes 18 years is the cutoff). The shows may require a minimum age, too. The exhibitors almost always have to own the cattle they're showing.

Shows are divided into classes so your animal competes with similar cattle. The most important distinction is breed. And within the breeds, heifers and bulls are judged in classes with animals of comparable age (usually within 30–60 days). Steers and market-type (not breeding stock) heifers are divided into classes by weight.

Classes for heifers are available at almost all shows. Steers, however, are more commonly shown at junior shows as opposed to open shows. And a few junior shows have classes for bulls, but they're primarily exhibited at open shows. The class winners advance to compete for breed or division champion. From those champions, the judge selects his two favorites as grand and reserve grand champion.

A wide variety of show opportunities exist, including

- **Jackpot or preview shows:** These shows, which may be open or junior shows, usually kick off the beginning of the show season. Depending on your geography, the season could start in the fall, winter, or spring. Because most of the shows are for young cattle just entering their show careers, they're a good chance for you and your animal to gain experience and assess your level of competitiveness.

These shows are usually scheduled so you can come to the fairgrounds, prepare your animal, show, and go home all in one day (or at the most be gone overnight). This condensed schedule helps cut down on your expenses and time commitment. These shows usually have classes for steers and heifers, but not usually bulls.

- **Expos or beef congresses:** These events are often the same time of year as jackpot or preview shows, but they take place over a few days. In addition to junior and open cattle shows, youth can often compete in showmanship and judging contests at these events. They have classes for steers, heifers, and sometimes bulls.

  Livestock equipment vendors are also often on site so you can do some comparison shopping. Consignment sales are frequently held in conjunction with these expos. They're a great opportunity for small or new livestock breeders to sell a few of their cattle in front of a large audience.

- **County fairs:** County fairs are most often associated with youth shows for participants in 4-H or FFA and are held during the summer or early fall. However, some county fairs offer a separate show open to anyone. The level of competition at county fairs can range from mild to quite intense. It all depends on the number of exhibitors and their level of expertise. County fairs often last for four to five days with the event culminating with the livestock auction.

- **Junior shows sponsored by breed associations:** At the national level, these shows offer a week-long immersion in everything cattle. Competitors from across the country show hundreds or thousands of cattle. But these events are so much more than just a cattle show. Youth can compete in other contests as well, including the following:

  - **Showmanship:** Exhibitors are evaluated on their ability to present an animal in a flattering fashion.
  - **Judging contest:** Participants rank numerous classes of cattle based on their conformation and give oral reasons for the placing.
  - **Fitting contests:** Youth work as a team to groom an animal for the show ring.
  - **Public speaking contests:** Contestants give speeches (either prepared or impromptu) on various aspects of the cattle industry.
  - **Beef knowledge quizzes:** Youth take tests to assess their knowledge of all things cattle.
  - **Photography, scrapbooking, beef-cooking, and advertising contests:** Competitors have a chance to demonstrate their creative side in these contests.

These summertime events are great ways for kids to learn, have fun, and develop skills that last a lifetime. We should know because between the two of us, we went to almost 20 of them!

If you're not ready for the time and travel commitment of a national event, several breed associations have lower-key regional competitions.

- **State fairs:** What's not to love about state fairs, with all the animals, the camaraderie, and the unique fair food (deep-fried Twinkie on a stick, anyone)? Junior shows at state fairs are often quite competitive, with top exhibitors from around the state vying for coveted champion banners and a spot in the potentially lucrative sale-of-champions. Open shows usually have fewer participants but are a good way for beginning seed-stock breeders to test the competitive waters and start to establish a name for their businesses. Even if you don't exhibit, attending the fair is a good way to network.
- **Major livestock exhibitions:** When it comes to open cattle shows, major livestock exhibitions are the big-time, and the competition in the junior portion is pretty stout, too. If you want to see the top show cattle in the country and watch how the professionals groom and present their cattle, visit one of these shows. Numerous breed sales take place during these exhibitions as well. Here are some of the biggest of these premier events:
    - **The North American International Livestock Exposition:** Held in Louisville, Kentucky, every November, this is the world's largest purebred cattle exhibition.
    - **The National Western Stock Show:** This show takes place in January in Denver, Colorado. Nearly 20 breeds are exhibited there.
    - **The Houston Livestock Show and Rodeo:** Approximately 2 million people attend this annual Texas show, which takes place in late February and early March.

You can find out about all the different types of shows from the following sources:

- Local and regional agricultural newspapers
- Your state beef association
- Your county extension office
- The national breed office for your breed
- Show websites

When you're looking for potential shows, be sure to check out the rules regarding exhibitor age, ownership requirements, and deadlines for when you must have possession of the animal.

Training your animal to be well-mannered in the show ring

A successful show experience depends in great part on the disposition of your animals. And disposition often depends on the selection of a well-mannered bovine. After all, your animal may have the world's best conformation, but you're not going to win any prizes if it's too wild to get to the show ring! Chapter 7 discusses the behaviors to look for and avoid when it comes to cattle disposition. In the following sections, we show you how to go about taming your calf for shows and exhibitions.

Getting to know you: Becoming comfortable with one another

Getting your animal accustomed to you is one of the most important steps in the training process. You want to start as soon as possible because the smaller the calf, the easier the task.

From the start, make sure that all your interactions with your cattle are positive. Always speak quietly and move slowly around the animal.

The ideal situation for taming a show animal is to house it in a small (approximately 12-x-10-foot) pen with access to a pasture lot. Keep the animal in the indoor pen for a few days while it settles down and becomes accustomed to being in close quarters with you.

Feed your show animal twice a day, and stay in the pen with it while it eats; doing so helps the calf make the correlation between you and the good stuff (like food). As it gets used to you, stand directly by the feed so the animal has to be close to you to eat.

Familiarizing your cattle with a show stick

Be sure to get your animal accustomed to a show stick. A *show stick* looks like a long golf club with a dull-pointed end and a slightly curved hook about 1 inch from the end of the stick.

Take the show stick into the pen with you at feeding time so the animal can see and smell it. You can use the stick to scratch the calf on its back, on the front part of its belly, or under its brisket and neck. Doing so keeps the animal calm and relaxed. The animal may be a bit surprised the first time you scratch it, so you may want to start by letting it smell the show stick and then gently rubbing the curved end of the show stick on the brisket, because that's often the itchiest spot on cattle.

You can also use the show stick to position the calf's feet so it stands in a nice-looking pose. Use the dull-pointed end to poke the front of the leg just above the hoof to move the leg backward. Use the hooked piece to grab under the dewclaw to pull the leg forward.

Introducing your cattle to a halter

Halters are an important tool for the show-trotting bovine. You use the halter to control your animal when you're leading it around the farm, fairgrounds, or in the show arena. After you have the animal quieted down by leaving it in a small pen for a few days (see the preceding section), move right on to *halter-breaking,* which is the process of teaching your animal to respect and obey a halter.

To train your animal to wear a halter, use a breaking halter. A *breaking halter* doesn't have a lead strap; the absence of this strap allows the cattle to wear the halter all the time without tripping. You can clip a lead rope to the breaking halter when you need one. The breaking halter is also a good choice for training because the part that goes under the chin is made of chain, making it easy to increase or decrease pressure depending on the animal's behavior.

After the animals are trained to calmly wear a halter and follow commands, you may want to switch to a rope halter, which weighs less and is more attractive than a breaking halter. A *rope halter* fits around the head snuggly (meaning that you can't increase or decrease pressure on the animal) and has a 4- to 5-foot lead strap attached to it.

To get the halter on the animal for the first few training sessions, follow these steps:

1. **Loop the part of the halter that goes behind the ear and over the poll onto your show stick. Use the show stick to guide the halter behind the poll and behind the ear farthest away from you. When the halter is in place, remove the show stick.**

   Throughout the process speak quietly and gently to your animal to help it stay comfortable. And if you haven't introduced your calf to a show stick, check out the earlier section "Getting to know you: Becoming comfortable with one another."

2. **With the calf standing quietly and turned toward you, hook the halter with the show stick again and bring the lower part of the halter up and around the muzzle.**

   For practice you can try these techniques on an animal that's already trained or even on a big fence post.

3. **Place the halter over the ear nearest to you.**

   Use the show stick if you can't get close; use your hand if your animal is calm enough.

Getting a halter on your animal isn't a timed competition. Don't be in a rush. Your calf will be more likely to stay calm and quiet if you do too.

A properly fitted halter should lay behind the ears and poll and come down the sides of the face a few inches below the eye. The part that goes around the lower head should be about halfway between the eyes and the tip of the nose. The lead strap should dangle from the left side of the animal's head. Figure 14-1 shows an animal with a correctly placed halter.

Start off with the halter fitting loosely because it's much easier to get on this way than when it fits snuggly.

To get the animal to follow you, use the lead strap of the rope halter or a lead rope clipped onto the end ring of the chain portion of the breaking halter. When the calf moves in the direction you want it to, let up on the pressure under the chin. When you let up on the pressure, you signal to the calf that it's behaving properly.

Start off leading the calf short distances. For instance, start off leading the calf from one side of the pen to the other, where you've left a feed treat. As your calf progresses in its training, you can shut it away from water overnight and then lead it to water first thing in the morning.

Figure 14-1: A correctly placed halter.

Teaching your animal to tie

You want to train your future show animal to tie. No, we don't mean it needs to be able to tie a bow. *To tie* means to stand quietly when tied by its halter to a post or fence. Your animal needs to stand peacefully while tied so you can groom it more easily. Cattle are also tied up when they are on exhibition at the fair. You can introduce tying right along with, or even before, you train your calf to follow a lead. Tie the calf with the lead strap to a secure post at a height about level with its chin. Leave about 18 inches of rope slack between the calf and the post. This amount of slack gives it space to move but not enough to get tangled up.

Keep your initial tying session to no more than 10 to 15 minutes long. By keeping the sessions brief, the animal is less likely to get restless or agitated. This is a good time to brush him and practice with the show stick (which we discuss earlier). You can read more about grooming in the later section "Shave and a haircut, two bits: Grooming your animal."

Progressing to advanced training

As your animal becomes better trained, expose it to different situations it may encounter away from home. For instance, consider doing the following:

- **Lead your cattle up and down the driveway.** This helps your cattle become comfortable obeying you and the halter in an area that's more open than their barn or pen.
- **Work them with the show stick both inside and outside of the barn.** Doing so trains the cattle to respond to the show stick even if their surroundings are unfamiliar.
- **Have different people walk around them and talk loudly.** The more people your animals are used to, the easier it'll be for them to adjust to all the new sights and sounds when you go to the fair.
- **Have an unfamiliar person role-play the part of the judge.** Doing so helps the cattle get used to having a new person stand close to them and touch them. (Judges often run their hands over a steer's ribs to check the amount of fat cover.)

Packing your feeding and grooming supplies for the show

By the time you finish packing for a show, you may feel like you're the butler for a movie star diva instead of the caretaker of a simple bovine. However, it's important to bring the supplies so you're prepared to make your animal feel and look its best at the show.

The following checklist can help you remember all the critical supplies you need to pack up for your cattle:

- **Bedding and tools:** Wood chips or bark are often used for bedding away from home. Straw is also an option. To maintain the bedding, take a pitchfork, rake, shovel, and broom.
- **Feed:** Take all the feed (both grain and hay) your animal is used to eating. Pack a little extra in case of a spill or some other accident.
- **Feeding and watering supplies:** Take feed pans for every animal and at least one water bucket for every three to four head. You also need a hose for filling water buckets and washing cattle.

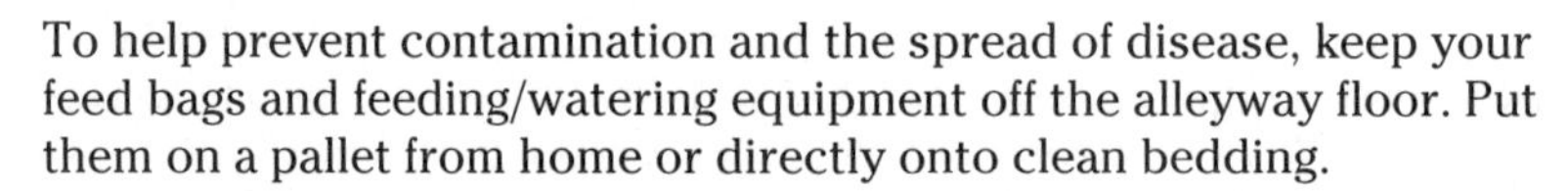

To help prevent contamination and the spread of disease, keep your feed bags and feeding/watering equipment off the alleyway floor. Put them on a pallet from home or directly onto clean bedding.

- **Grooming chute:** A grooming chute comes in handy to hold your animals still for the fitting process.
- **Hair products:** Bring whatever hair products you normally use, including clippers, scotch combs, rice root brushes, blowers, spray adhesives, curry combs, and wash soap.
- **Halters and neck ropes:** You need both regular rope halters and leather show halters. The neck ropes clip around the neck and are a backup in case an animal slips its halter off.
- **Miscellaneous:** Don't forget these various miscellaneous items: Show harness (to hold your exhibitor number) and show stick, stall cards, first-aid kit (see Chapter 9 for details), and extra tools and supplies, such as wire, duct tape, and a Leatherman multi-tool.
- **Paperwork:** Assorted bits of paperwork are necessary for cattle on the go. Be sure to bring completed entry forms or receipts, health certificates from the vet, registration papers, and business cards.

Making the Most of the Show

After you've prepared for months for a show, you want to do all you can to put your farm and your cattle in their best position to shine when you finally arrive. To make the most of your exhibition experience, present yourself and your animal in a professional manner, keep your animal nicely groomed, and use proper show ring etiquette. We show you how to do all these things in the following sections.

Putting your best foot (or hoof) forward

The image you, your animals, and your barn portray has a big impact on what potential customers may think about your business. Even though caring for cattle can be a dirty job, dress neatly in good jeans or dark slacks with boots or sturdy work shoes and a clean, pressed shirt. It's a good idea to have an extra set of clothes to change into after the morning feeding and washing is done, especially on show day.

Having your name and farm name embroidered on your shirt allows other people to easily identify you and eliminates that uncomfortable situation where folks keep saying, "I know you, but I forgot your name."

To keep your cattle on their regular routine and looking good, get to the barn early so your cattle can eat breakfast at their normal time. On days leading up to the show, you can wash your cattle and dry their hair after they have eaten breakfast. On show day, however, reverse the order so you have more time to get their hair completely dry. When your cattle stand up from resting, brush their hair to remove any bedding. Also use a pitchfork to pick up and remove any manure.

Keep your stalls neat and clean and the alleyway nicely swept. Bring only freshly painted equipment, and store it neatly. Post easy-to-read signs and business cards with your farm name and contact information. Also put stall cards above each of the animals you're exhibiting. On the cards, list, at minimum, the following information: animal name, birth date, sire, and dam.

Shave and a haircut, two bits: Grooming your animal

The purpose of grooming your animal is to accentuate its positive physical characteristics and hide its flaws. Grooming your animal at the show can be a fun, yet intense, experience.

To take some of the pressure off yourself and to get your calf accustomed to all the grooming-time fuss, do at least one trial run at home. Doing so gives you a chance to make sure you have all the needed supplies and that your equipment is working properly.

Time yourself during your practice run at home so you have an idea of how long you need to groom your animal before the show. You don't want to be too rushed nor do you want your animal to get tired out waiting around in the grooming chute. We give you the lowdown on trimming, shampooing, brushing, and primping your animal's hair in the following sections.

Trimming

You need to give bovine hair at least one — if not two — trimmings with clippers to get it looking its best and blending smoothly over the body. Clip the hair on the head (except the poll), neck, and brisket very short. However, be sure to taper these areas so they blend in with the hair on the shoulder and don't make abrupt changes from short to long.

Trim the hair on the body so it's as fluffy and full as it can be without appearing long and scraggly. The exception is cattle with short summer coats or cattle with short Brahman-type hair. In these cases, your best bet is simply to trim and brush the hair, so the animal looks clean and shiny.

Shampooing, brushing, and primping

After trimming your cattle's hair (see the preceding section), you're ready to shampoo, brush, and primp for the show. You want to wash your animal thoroughly the day before the show. You can use special cattle hair soap if your cattle are really dirty; however, a mild liquid dish soap works fine for cleaner animals.

The day of the show, do a quick rinse, soaping up only the spots that are really soiled. Doing so saves you drying time because the hair will be only slightly wet rather than completely saturated from a thorough scrub and wash. This extra time is especially important if you're in the first class of the morning. If you have the time, it's fine to do a complete wash and blow-dry the day of the event.

When blowing the hair dry, keep the blower nozzle pointed at a 45-degree angle to the body so the hair stays fluffy and lays in the proper direction (pointed toward the head at an upward angle). After blow drying, use a rice root brush to brush the hair in a forward direction from the tail to the head. Besides ensuring the hair is laying properly, brushing from back to front helps remove dead, itchy skin and gives the hair a healthy shine.

For the final preparations, lead your bovine into the grooming chute and tie its halter to the front of the chute. Then follow these steps:

1. **If the animal has hair at least 1-inch long, apply a mousse-type adhesive to the coat.**

   Use a blower on a low setting and a scotch comb to get the hair in the desired position.

2. **Apply a misting of adhesive spray to hold the hair in place.**

   Start misting down by the back hooves, and work your way up and forward so you're always working the hair in the desired direction.

3. **Use clippers to shape the hair that is held in place by the adhesives. This final trimming and clipping is in addition to the basic haircut you already gave your animal.**

   Give the animal a level-looking back and a smooth, trim head, neck, and belly. Don't take off too much hair on the sides and rump; you want the animal to look full and muscular. On the legs, trim the hair only a little so the bone looks good and strong. Trim the area around the leg joints a little closer to keep them from looking swollen.

4. **Slightly tease and rat the tail switch.**

   Teasing the switch makes it fuller and helps fill in the space between the hind legs to give the calf a meatier appearance.

5. **Spray the animal all over with a product designed specifically to give the hair extra shine (look for sprays with names like "final bloom" or "show sheen").**

   If show rules allow, you can also use special cattle hair spray paint to make the hair a uniform color throughout.

6. **Apply adhesive spray to the hair on the poll, and pull it up into a peak to make the head look sleeker and longer.**

7. **Replace the rope halter with a clean and shiny leather show halter.**

When the show is over, you have to apply an adhesive-removing product all over the animal and rub it down. Follow that up with a good soapy wash (use a special soap designed for gluey, sticky cattle hair) and thorough rinse.

Following show ring protocol

Here comes the moment you have worked so hard and waited so long for: It's time to show your animal! Keep the following tips in mind before you get in the ring:

- Unless you're the first class of the day, take the time to watch another class show so you can see how the judge wants the cattle positioned in the ring.
- Be sure to have a *show order* (a listing of all the classes and animals in each class) in your stalls to keep track of when it's your turn.
- Scope out the arena before the show starts so you know where the high and low spots are located. Cattle look best if their front feet are on a slight rise. Avoid putting their front feet in a hole; they'll look quite awkward.

The following list of steps gives you an idea of how to proceed when it's your turn to make your way into the ring and into position for judging:

1. **Arrive at the *make-up ring* (the waiting area adjacent to the show ring) promptly when called so your animal has time to take in the crowds and noise.**
2. **Watch for the *ring steward* (the person who helps direct the cattle and exhibitors around the ring) to signal to you to enter the show ring.**

   Keep an animal-length distance between you and the next animal when walking into and around the ring.
3. **When walking around the ring, keep your show stick in your left hand so it's positioned with the pointy end directed toward the ground. Hold the lead strap in your right hand.**
4. **When you reach the judge, pull your animal straight in beside the next one (not at an angle) so the judge can walk down the row of animals and examine their heads and then move around to look at their tail ends.**
5. **As you come to a stop in position, switch the lead strap into your left hand and the show stick into your right. Work quickly yet calmly to get the animal's feet positioned with the show stick.**

While you're in the show ring, pay attention. You need to be aware of not only your animal but also of the other animals and exhibitors, the judge, and the ring steward. Many times the judge uses hand signals to direct you into position.

After looking at the animals side by side, the judge will have you walk them around the ring. When the judge is watching your animal walk, make sure it's moving at a pace that best shows off its ability. Keep its head pointed forward so its whole body moves in a straight line. After the judge has seen all the animals walk, stop and use the show stick to set up your animal in a head-to-tail, or profile, lineup.

Keep your eye on the judge; she will usually pull the cattle into at least an initial, if not final, ranked order after looking at the animals in profile. You don't want to miss the directive to pull into first place!

Figure 14-2 shows how cattle should be positioned in a head-to-tail lineup. In this figure, you can also see the proper use of a show stick.

Figure 14-2: Cattle in the show arena.

Exploring Sales Opportunities for Your Cattle

Raising cattle is fun and rewarding in its own right, but unless you have an endless supply of money and land, you eventually need to sell some of your herd. In the following sections, we take a look at some different selling options.

Opting for private treaty deals

When buyers purchase cattle directly from you and not through a public sale, this transaction is referred to as a *private treaty* deal. By working directly with the buyers, you can set the price and don't have to worry about meeting any third-party requirements (other than those required by state and federal animal health laws). You get to keep the full purchase price, and you don't have to pay any entry fees or commissions. Having customers come to your farm is also a great way to showcase your entire program and hopefully develop long-term relationships with your customers.

With private treaty deals, you need to put forth the effort and expense of getting people to your farm by advertising and participating in shows and sales. You also have to prepare yourself for the possibility that potential customers may come and take up most of your day looking and then leave without buying anything.

Participating in consignment sales

Consignment sales, where multiple cattle breeders enter or consign animals to a public auction, can be an excellent way to sell cattle. These sales are often sponsored by a breed association or state cattleman's organization. They also may be held in conjunction with cattle shows. These independent third parties usually require your cattle to meet performance or health requirements in order to be eligible to sell. The guidelines assure buyers that you have healthy, high-quality cattle. Depending on the sale, consignment offerings can include herd bulls, bred cows, cow/calf pairs, and show heifers or steers.

Because multiple cattle breeders enter animals in these sales, the costs of advertising and conducting the sales are spread out and reduced compared to putting on a sale of your own. In fact, if you're a new breeder, you can actually let the reputation of the sale pull the buyers in, and then you can let the merits of your cattle speak for themselves.

Costs for consignment sales include the entry fee to participate and a commission based on the sale price of your animal. You also must factor in the time, cost, and effort involved in getting your animal prepared for and transported to the sale.

If you're raising commercial feeder cattle, look into special consignment sales at your local auction barn. As opposed to the regular weekly sale where all types of cattle and other livestock sell, the special sales offer only feeder calves. These sales usually attract more buyers, and they're often willing to pay more than buyers at regular sales for calves that are documented as being weaned and/or vaccinated against disease.

Getting a hoof in the door by enrolling in a performance test

A *test station* is a program, usually run by the state extension service or land-grant university, where bulls from various farms come to one location to go on *test.* The test is basically an assessment of the animal's performance over a 90- to 150-day period. All the bulls are fed the same diet and measured for their growth, *feed efficiency* (how much feed is needed to gain a pound), reproductive soundness, and carcass merit (via ultrasound). After completing the test, the higher-performing bulls (usually the top half or two-thirds) are offered for sale through a public auction.

You do have to pay an entry fee for each bull you put on test to cover its feed expenses, preventive healthcare, and other costs associated with running the test and sale. However, keep in mind that tests are a viable way for small (and large) producers to sell bulls. It can be difficult for the small producers to justify enough advertising to adequately promote and sell just a few bulls. But by enrolling your bulls in a performance test, you can take advantage of the notoriety and creditability associated with the test. Additionally, you benefit from getting valuable performance information, such as carcass data or feed efficiency, that would be difficult to collect otherwise. Plus, if your bulls perform well at the test and for their future owners, you build a positive reputation for your cattle and your breeding program.

Selling your cattle online

The Internet has become a powerful merchandising tool in the cattle industry. It can help you reach customers all across the country without the time and expense of traveling and consigning to sales. You can use the web in several different ways to sell all different types and numbers of cattle. Here are the most common:

- **Create a farm website.** You can build your own website using free downloadable software and have your pages hosted for less than $10 a month. Your site can be a place to share your breeding program's philosophy and goals. You can post pictures of your herd sires and top dams. Include photos and maybe even videos of the cattle you're selling. If you have show winnings or performance data, share that information, too. Highlight any sales or shows you'll be attending so customers know where to find you.
- **Post to online classifieds.** Numerous state and regional agricultural publications offer online classifieds along with their print advertising. You can also post your cattle for sale on some general sites for free or pay to advertise on more cattle-specific listings.
- **Check out breed association websites.** Most breeding associations maintain a website and offer their membership advertising privileges. Using these resources is a good way to reach out to those customers who are searching for a particular breed of cattle.

Strike a pose: Taking photos of your cattle

When it comes to merchandising your cattle business, a good picture is worth a thousand words. Here are a few tips to help you get quality, eye-catching photos of your cattle:

- **Keep distractions to a minimum.** Doing so helps the focus of the picture be on your animal, not an old pile of junk in the background. Also you want your animal to be at ease in the photo environment, not nervous and startled by unexpected noises or movements.
- **Take pictures after the cattle have eaten.** Don't try to get candid shots at feed time unless you want a bunch of animals with their heads down eating. Enter the pasture quietly after the cattle have been grazing for awhile so they have full bellies and won't be looking to you for feed.
- **Have another person help you if possible.** Cattle look best when their ears are perked forward, their head and neck are angled 45 degrees upward from their shoulders, and they aren't looking directly at the camera. So when photographing tame cattle, ask a person to stand out of the picture boundary and make strange noises or shake a hat to get the animal's attention.
- **Don't take photos in the middle of the day.** When the sun is directly overhead, it casts unflattering shadows on the animal's body. Early morning or late afternoon are the best times to take photos. Shooting on cloudy or hazy days also helps minimize shadows.
- **Take the photo with your animal's front feet standing slightly uphill from the hind feet.** Never take pictures of your cattle stand downhill; even an award-winning champion looks bad in that position.
- **Get down on the ground for pictures.** You want the camera to be level with the belly of the animal. Take the photo straight from the side not angled from the front or rear. This position keeps the animal in proper perspective, allowing it to not appear abnormal in size.

Chapter 15

Managing Your Beef Business

In This Chapter

- Checking out the regulations for selling beef
- Determining your products and prices
- Working with your meat cutter
- Selling and promoting your beef

Producing food for people is a gratifying endeavor, but it's not always an easy one. Your responsibilities grow beyond the challenges of animal husbandry tasks to food quality and safety control, working with a butcher, complying with government requirements, and finding and retaining customers. However, don't let the extra work keep you from raising and selling beef direct to customers. Many hungry consumers are waiting to make a real connection with the people who produce the food they eat and serve to their families.

In this chapter, we cover what it takes to go from cattle producer to beef producer. We provide information about what you need to consider from a legal and liability standpoint, and we help you understand what different cuts of meat a beef animal produces and how you can price those products.

Having a good meat cutter is also critical to your success in the beef business. So we talk about what to look for when selecting this important partner. As you do all the background work involved in selling beef, you also need to work on promotion. This chapter gives you ideas about different marketing venues and things you can do to sell your product.

Covering the Legalities of Selling Beef

After you transition from raising animals to producing and selling food, you encounter numerous laws and regulations. To protect yourself from fines, or worse, you need to follow the rules and have adequate insurance coverage. We discuss all the important legal topics — facilities, paperwork, and insurance — in the following sections.

Understanding governmental rules regarding processing facilities

Meat can be processed in several types of locations. The meat butchered at each type of facility has varying regulations regarding to whom the meat can and can't be sold to. Here are the most common processing facilities you have to choose from, depending on your needs and scope of business:

- **Facilities inspected by the United States Department of Agriculture (USDA):** These facilities are also known as *federally inspected meat processing plants.* Meat processed at these facilities can be sold across the United States and may be eligible for shipment to foreign countries.

  About half the states in the country have only federally inspected plants. Many of these facilities are owned by a large company and won't butcher livestock not owned by the company.

- **State-inspected facilities:** Federal government requires the level of inspection at the state facilities to be "at least equal to" the level of federal inspection. State-inspected facilities are often family-owned and willing to work with small producers to get their animals processed. The selling options for meat processed in a state-inspected plant vary depending on the state. Some states allow processors that have worked with the USDA to sell their products across state lines. In other states, meat butchered in a state-inspected facility can only be sold within that state. Still other states don't permit meat from a state-inspected plant to be sold at all.

  The legal selling status of meat from state-inspected facilities is constantly changing. Contact your state department of agriculture or your state board of animal health for up-to-date regulations for your location.

  State-inspected processors can be found in about half the states in the country. For a detailed list go to www.fsis.usda.gov/regulations_&_policies/Listing_of_Participating_States/index.asp.

- **Custom-exempt facilities:** The meat processed in custom-exempt facilities can't be sold and is labeled as "not for sale." These facilities don't undergo a routine daily inspection process by a government official. However, this status doesn't mean these facilities aren't safe for meat processing. The producer of an animal can have it butchered at a custom-exempt facility, and then he can use the meat for his own personal use, for family use, for nonpaying guests, and, in some states, for the producer's employees.

  Some states stipulate that you can't buy an animal and then have it processed at a custom-exempt plant. Other states do allow farmers to sell an animal to a buyer who can then have the animal slaughtered at a custom-exempt facility. Check with your government authorities for the specifics in your state.

- **On-farm processing:** You can process beef yourself on your farm for your own personal consumption. These beef products can't be sold to customers.
- **Mobile processing units:** Another great, although uncommon, option is a state- or federally inspected, mobile, on-farm processing unit. A few of these roving facilities exist, and they actually go to the farm and do the processing onsite. Ask your beef extension agent whether any of these mobile units are in your area.

Regardless of where your beef is processed, it is your responsibility to obey proper medication withdrawal. *Withdrawal time* is the number of days that must pass between when an animal last receives medical treatment and when the animal's meat is approved for human consumption. You must also obey these withdrawal rules when selling live animals through auction.

Obtaining permits and licenses

You must obtain and organize all the necessary paperwork needed to sell meat. You need permits and licenses so you're in compliance with government laws and don't run the risk of fines or having your business shut down. You also want to have all your documentation in order in case you need it to file an insurance claim.

Here's a rundown of the types of permits you may need:

- **Food safety and inspection permits:** These may include a copy of the permit issued to your butcher from the state or federal meat inspection agency, a health permit (issued by the local or county health department) for every county in which you do business, and certification showing at least one person in your business has been trained on proper food handling techniques (one common certification program is ServSafe). Contact your local extension service about food handling classes and certification opportunities. (To find contact information for your local extension agent, go to `www.csrees.usda.gov/Extension/USA-text.html`.)

  Any labels you put on your packages of meat will also need approval from the meat inspecting agencies. Your butcher and local meat inspector can direct you to the proper forms to fill out for this process.
- **Business and tax permits:** For most states you need to register your business with the state or local government and receive a certificate or license recognizing your business as a retail merchant. Your local chamber of commerce or economic development office can help you navigate the requirements specific to your geography. Depending on your local laws you may need to have a mechanism in place to collect sales tax on the products you sell.

Don't be discouraged by the potential bureaucratic tangle. If you're planning to participate at a farmers' market, the market master can often clue you in on local regulations or can at least tell you the proper government office to contact.

Protecting yourself with proper insurance coverage

Insurance is one of those things people buy but hope to never need or use! You should consider several areas of insurance protection for your cattle and beef business:

- **Farm policy:** This policy covers damage caused by the cattle. It may include coverage for the following:
  - Crops that are damaged when your cattle get into your neighbor's soybean field
  - A vehicle and bovine collision
  - Reimbursement for the loss of cattle due to a barn fire or lightning

  These are just a few of the coverage possibilities. Check with the provider for a complete list of coverage.
- **Location liability coverage:** If you're going to have customers come to your farm and buy product, or if you're going out in public to sell meat (like at a fair or farmers' market), get this coverage. It protects against injuries that may occur when customers are at your booth or stand.
- **Product liability coverage:** If you sell meat, purchasing this type of coverage is very important. This policy provides coverage if consumers have a food safety problem with your beef.

Exploring and Pricing the Meat a Market Animal Produces

Before you can knowledgeably answer customer questions and sell beef, you need to bone up on some basic information about the amount and types of meat a beef animal produces. And, unless you're an altruistic producer who isn't concerned with making money (or at least breaking even), you need to educate yourself on how to fairly price your product. In this section, we take a look at the different forms in which you can sell your product and how you can be compensated.

Selling the whole animal on the hoof

A straightforward way to sell beef is to not sell beef at all! What we mean is that instead of selling a customer meat, you sell a live animal to another individual or family (who then has the animal butchered for meat). This transaction is called *selling on the hoof.* If you're raising a few head to market weight to fill your own freezer and have an extra animal, selling on the hoof may be an option for you.

When selling on the hoof, usually you transport the animal to the butcher shop where it's weighed and processed. You figure the price of the animal by multiplying its weight while alive by an agreed-upon amount per pound (see "Setting your prices" later in this section for help determining a fair price). A beef animal ready for butchering may weigh as little as 950 pounds or as much as 1,400 pounds. (Chapter 13 discusses how to determine whether your cattle are ready for butchering.) The customer should pay you for the beef after it's weighed and before they pick up the meat from the butcher shop. Talk with your customer about what size animal they want and how lean or fat it needs to be. The consumer is responsible for working with the meat shop regarding how the beef is cut and packaged, and then he pays the butcher directly for the processing.

Providing freezer beef

Many consumers don't have the need for a whole beef, nor do they have an interest in buying a live animal. Plus they may not have the knowledge to work with the butcher regarding how the meat should be cut. You can provide a needed service by facilitating this transition from steer in the pasture to meat in the freezer. *Freezer beef* is meat that is bought frozen and in bulk. It can be as little as a 20-pound variety of steaks, roasts, and ground meat up to the beef produced by an entire animal (300 to 550 pounds).

If you're going to sell beef and not the actual animal, you have to have the processing done at a federally-inspected or, depending on your local laws, a state-inspected facility. (To read about these facilities, refer to the earlier section "Understanding governmental rules regarding processing facilities.")

When providing freezer beef, you have to determine how you'll have the butcher cut the meat and also how to price it. We explain both in the following sections.

Choosing how to have the meat cut

You can have all your freezer beef cut the same way so the customer doesn't have any options, or you can serve as a go-between to educate the buyer on all their different cutting and wrapping choices and then convey that information to the butcher.

A freezer beef can be split up in a variety of ways; here are some of the most popular:

- **Side of beef or ½ beef:** The customer gets half of the meat produced from a beef animal, either the right side or left side.

  The animal only has one tongue, oxtail, or hanger steak, so be sure you don't promise these cuts to each buyer.
- **Front-quarter or hindquarter:** The customer gets only meat that comes from the front part of the animal (chuck, rib, brisket, shank) or the back (loin, sirloin, round, flank). Because the back quarter has more of the higher value steaks, its price is usually more per pound than the front quarter.
- **Mixed quarter or split side:** The customer receives 25 percent of the beef an animal produces, so she gets a combination of the cuts from both the front-quarter and hindquarter.

Figure 15-1 can help you visualize the way a beef carcass is broken apart and all the cuts that come from it.

Pricing your freezer beef

You can set the price for freezer beef based on the animal's live weight (as we discuss earlier for selling on the hoof), by the carcass weight, or by the pounds of meat produced. All three pricing methods are commonly used.

Your customer can pay the butcher directly, or you can include the cost of processing in your price and pay the butcher yourself. When selling freezer beef, it's a good idea to have the customer pay a deposit (usually $100 for a quarter or larger amounts of beef) and then pay the balance to you at pick up or delivery. We have our customers pay us, and then we settle with the meat cutter. Our customers seem to appreciate knowing that one price includes the meat and processing.

Marketing your meat by the individual cut

If you're interested in selling at farmers' markets or to restaurants, be prepared to sell your meat one cut at a time. Many folks don't have the need or the freezer space for a quarter or half beef. Similarly, restaurants rarely want the variety of cuts that come with a freezer beef; they're more interested in larger volumes of one or two particular cuts.

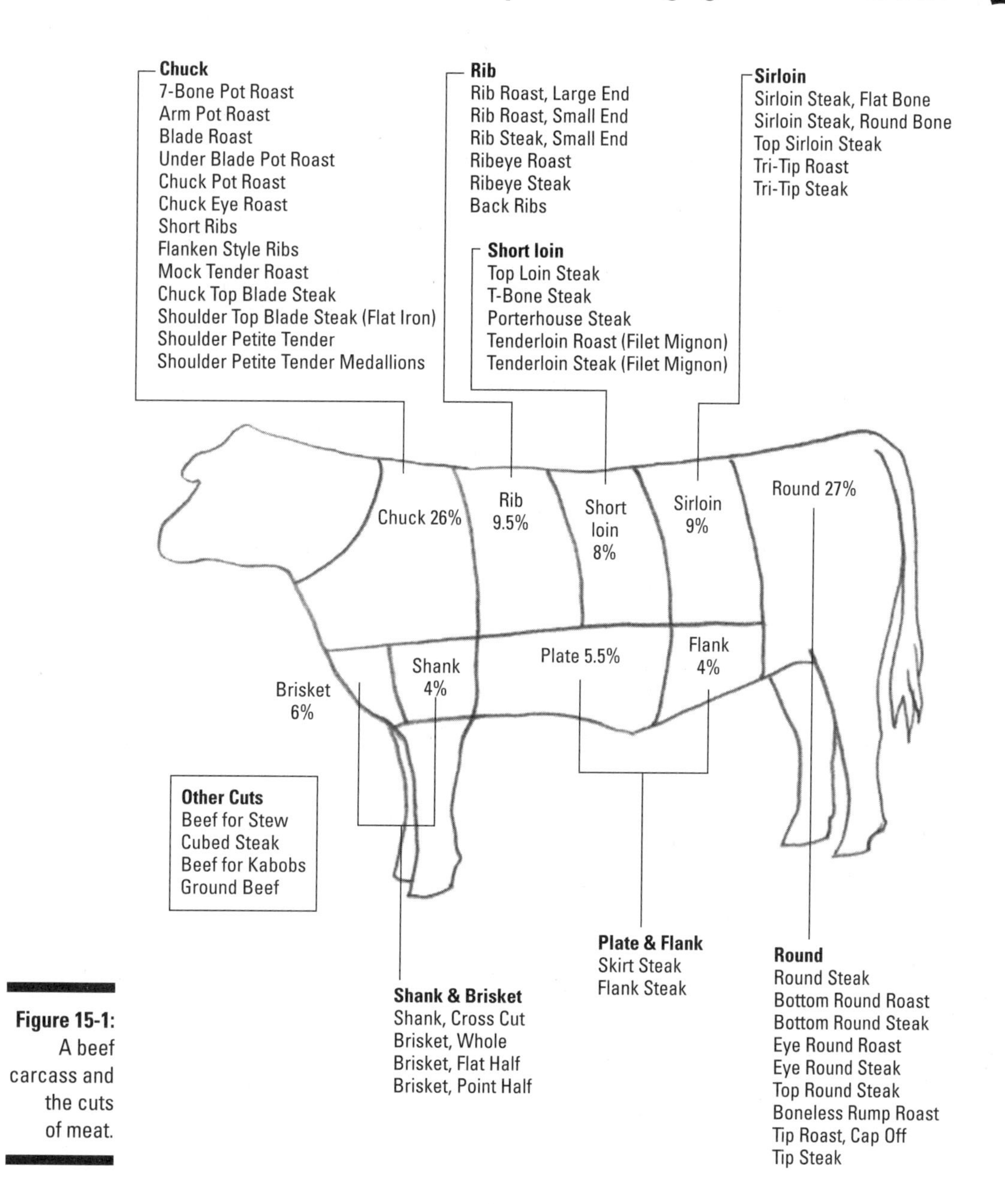

Figure 15-1: A beef carcass and the cuts of meat.

Depending on your clientele, you may sell out of steaks in a flash but be stuck with hundreds of pounds of ground beef. During the winter, you may run out of roasts and stew and have a backlog of steaks. The key to success when selling individual cuts is to keep your inventory in balance and to sell the entire beef. See the nearby sidebar "Making seasonal changes to our product lineup" for some of our solutions for balancing inventory.

Table 15-1 shows the approximate weight of each cut yielded by a 1,000-pound beef animal. All cuts are boneless unless noted. Understanding the amount of each cut an animal produces will help you determine how much you will have available to sell.

Table 15-1 Weights of Beef Cuts from a 1,000-Pound Animal

Cut	*Weight (in pounds)*
Ribeye steak	12
New York strip steak	11
Filet	2.5
Sirloin steak	12
Round steak	32
Bone-in chuck roast	50
Bone-in arm roast	20
Sirloin tip roast	12
Rump roast	10
Tri-tip roast	2.5
Stew	20
Brisket	4
Short rib	12
Ground beef	150
Flank	3
Skirt	4

Clearly label and price your beef sold by the package. Your butcher shop should have the appropriate rubber stamper or labels to identify each package of meat.

If you plan to weigh the meat yourself at a farmers' market, you need to have your scale inspected and approved at least yearly by a government weights/scales official.

Setting your prices

Few people are in the beef business just for pleasure. Everyone wants to make a profit — even if they love what they're doing. To make a profit, you have to price your products fairly but competitively. You can take two basic approaches to setting your prices:

- **Commodity pricing:** You can base the price of your animal on the value of the commodity cattle bought by large, national beef processors or cattle that have sold through your local auctions. You can find several sources of commodity cattle prices, including the following:
  - Agricultural newspaper publications or their websites
  - The USDA website at `www.ams.usda.gov`
  - Sales at your local auction barn
- **Your costs plus a markup:** Another pricing option is to determine how much money it takes to raise your cattle and what amount of profit you need to make and then figure your price from that data. Producers that sell average, nondescript cattle on the hoof (see the earlier section "Selling the whole animal on the hoof" for more on this term) often charge a few more pennies per pound than what commodity cattle are selling for, especially if the cattle have been raised in a low-labor, low-cost feedlot situation.

  However, as the attributes of your cattle increase, you should raise your price accordingly. Practices such as using organic feeds, raising cattle on pasture, producing a certain breed or type of cattle, and minimizing the use of hormones or antibiotics may increase the value of your product (and the expense in producing the beef), allowing you to adjust your prices accordingly.

No matter which approach you take to pricing, if you — not the ultimate customer — are paying the butcher, be sure to include the cost of processing in the final price you charge.

If you're selling based on the carcass weight or by the actual pounds of beef the customer takes home, figuring out pricing gets a little more complicated because it's harder to find comparable commodity prices for carcass and retail cuts than it is for live animals. You also need to make sure you're setting prices that are fair to you and your customer.

When it comes to determining carcass weight, be consistent and always use either hot carcass weight or cold carcass weight. *Hot carcass weight* is the weight of the carcass taken soon after the head, hide, and internal organs have been removed. *Cold carcass weight* is the weight of the carcass after it has chilled in the cooler. The cold carcass weight can be 10 percent less than the hot carcass weight due to evaporation of water from the carcass while it hangs in the cooler.

You need to increase the price customers pay for a carcass compared to a live animal. A general rule of thumb is the carcass should be priced at 160 percent of the live animal price. So if you would sell your 1,000-pound animal for $1 per pound, you should sell its carcass at $1.60 per pound.

Making seasonal changes to our product lineup

To help meet customer demand and keep our inventory under control, we modify how our beef are cut on almost a weekly basis. During grilling season, we divert roasts into ground beef patties. However, we grow our stockpile of stew meat and roasts in the late summer because the demand for those cuts goes way up after a few chilly fall nights. Around Thanksgiving and Christmas, instead of cutting the rib and filet into steaks, we leave those cuts as fancy roasts such as prime rib and tenderloin for special holiday meals. We also make value-added products like sausage or deli-style roast beef when we have an excess of ground beef or beef round, respectively.

Be sure to properly set customer expectations about what they get for the price. Most people are surprised the first time they buy a 1,000-pound beef animal and only take home around 360 pounds of meat. Even if you sell based on the carcass weight, tell your customers that a 600-pound carcass yields 58 to 60 percent take-home product.

Just as the price per pound increases as you move from selling a live animal to selling a carcass, the price continues to increase when you sell quarter, half, and whole beef by the pound. Determine how much each live animal is worth, and then divide this amount by the pounds of meat produced. The resulting figure is the price per pound you should charge for each pound of meat.

As you begin setting your prices for individual packages of meat, use Table 15-1 as a guide to help you know what amounts of various cuts to expect. However, definitely assess how several of your own beef cut out. Your cattle may yield better, or your butcher may cut the carcass differently.

Finding and Working with Your Butcher

Finding a good butcher is critical to the success of your beef business. All your hard work in raising healthy, high-quality cattle and marketing your beef won't be worth anything if you can't find a meat cutter that does excellent work.

Even though talented, conscientious meat processors do exist, be prepared to travel a long distance to find one. They're in high demand and low supply. Your state department of agriculture, board of animal health, or extension service may be able to provide you with a list of meat processors in your area. Look across state lines for federally-inspected facilities, too. (To find contact information for your local extension agent, go to `www.csrees.usda.gov/Extension/USA-text.html`.)

The following sections provide some tips on how to evaluate prospective meat cutters and how to develop a good working relationship with the one you ultimately choose.

Checking out the facilities and personnel

When you're on the hunt for a meat processor, you should visit facilities and determine whether they work under the same principals and standards as you. Here's a list of ways to evaluate a prospective facility:

- **Confirm the level of inspection the plant operates under, and determine whether that level is sufficient for your marketing plans.** For example, if you plan to sell your meat in other states, you need to have your cattle butchered at a facility that's approved for processing meat that's eligible for interstate shipment. You can read about inspections in the earlier section "Understanding governmental rules regarding processing facilities."
- **Talk with the owner and manager to see whether they're truly interested in doing business with you.** If they aren't positive and excited, they probably won't do a good job.
- **Make an appointment to visit the facility and take a tour.** Look at the *kill floor* (where the animals are killed and the heads, internal organs, and hide are removed), check out the coolers, and watch the meat cutters and wrappers at work to evaluate the overall cleanliness and skill of the facility and staff. Look at the packages of meat to see whether they're wrapped and labeled in secure and nice-looking packages. Make sure they have the freezer capability and space to quickly freeze your product (a fast, hard freeze keeps quality high and reduces the chance of freezer burn).
- **Gauge the adaptability of the personnel.** Are they willing to cut meat according to your specifications, or are they set on doing things "their way"?
- **Determine whether the butcher shop has the equipment to prepare specialty products like beef jerky, smoked/cured meats, and summer sausage.** Being able to offer these products is not absolutely necessary but it does help broaden your appeal to more customers and can aid in balancing inventory.
- **Check with the manager on the financial issues.** Get a price list of all the services the butcher offers. Confirm how your bill is calculated so you can correctly figure the prices you need to charge your customers.

Considering your animals' well-being

Of paramount importance is to understand and confirm how the butcher treats your animals. You need to be clear about your expectations regarding how your animals need to be treated. All animals deserve respect and proper care, and you want to have your animals processed by people who put those beliefs into practice.

Here are some general things to look for when evaluating a facility based on its animal welfare:

- **The butcher shop should have a clean, safe, sheltered holding area where the cattle have access to clean, fresh drinking water.** The area should also be as stress-free as possible. It should be quiet and free of distractions, such as shadows or random objects.
- **The cattle should be moved to the *stun box* (where they're rendered insensible to pain) in a quiet and calm manner.** The processor should be trained and skilled to efficiently stun cattle on the first application of the stunning device.
- **The processor should work quickly.** No set time exists from stunning to bleeding out (cutting of the arteries and veins to release blood from the body), but it should happen quickly so the animal doesn't regain or begin to regain sensibility during the process.

Deciding on packages and labels

Of course, the packaging and presentation of your products count for something. Not only does the packaging keep the meat fresh and protected, but it also suggests the level of quality. The main options and considerations for packaging your meat are the following:

- **A double layer consisting of white or brown wax-coated butcher paper and heavy-duty plastic wrap:** This packaging method has an old-fashioned appeal. It reminds people of going into the corner store, picking out the meat for dinner, and having it wrapped by the meat cutter just for them. The meat stays well-protected wrapped this way and isn't too susceptible to freezer burn.

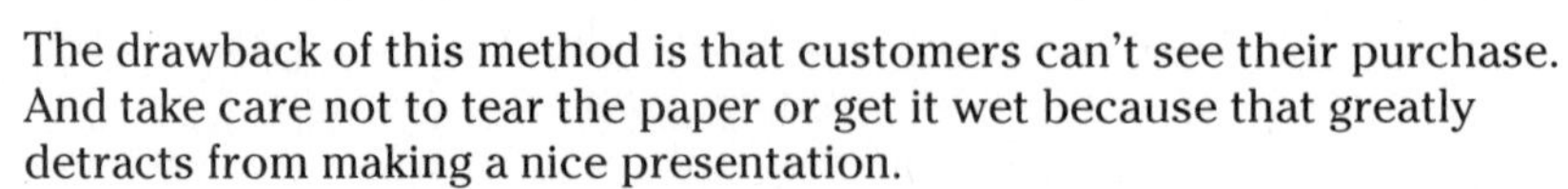

  The drawback of this method is that customers can't see their purchase. And take care not to tear the paper or get it wet because that greatly detracts from making a nice presentation.
- **Vacuum-sealed plastic packages:** The transparency of these packages allows the customer to see the meat. If you go this route for packaging, however, be sure the meat wrapper gets a good, tight seal around the meat. If air pockets pop up between the packaging and the meat, ice

crystals quickly form when the product hits the freezer. Because the meat is actually showing, be sure it's frozen in a nice shape; a frozen lumpy mass doesn't do much to increase appetites.

To finish off the package, consider using your own custom-made labels instead of the butcher shop's labels. They can give your packaging a distinctive and professional look. Have labels made bearing your farm name and information. Just be sure they fit into your butcher shop's scales and printer. The added expense of these labels is worth it as you work to build brand identity and spread the word about your products. Having your contact information on each package also gives the consumers confidence knowing they can reach you directly with questions or concerns.

Sometimes processors replace their name with yours on their in-house labels. This option is better than no farm name at all, but it can lead to confusion if other producers are using the same processors and, hence, the same labels.

You need to have all meat labels approved by a federal- or state-inspection agency. Approval can be a lengthy process, so plan ahead. Also be prepared to verify any marketing claims, such as "high levels of conjugated linoleic acid (CLA)" or "100 percent grass-fed."

Maintaining a good relationship with your chosen meat cutter

Treat your butcher nicely. (Just like mom said.) Not only is it the right thing to do, but the production of your beef depends on it. Unless you have the skill to cut your own beef and a spare $250,000 to build a butcher shop, you need to stay on good terms with someone — like your butcher — who does.

The following tips can help you make the most of the critical working partnership between you and your butcher:

- **Review your processing needs with your butcher in advance so she can plan for personnel and equipment needs.** This advanced notice is especially important if you'll be processing a large number of animals, if your butcher has many customers, or if she will be processing large amounts of wild game like deer or elk during certain times of the year.
- **Provide pictures or photos of how you want your different meat cuts to look.** Just be careful not offend anyone or imply that the current practices are wrong.
- **Supply clear and consistent cutting and wrapping instructions.** Typed directions are easier to read than handwritten notes. Put the directions in the same format, and highlight any deviations from your standard protocol.

- **Pick up your meat once it's frozen.** Don't let it linger in the meat processor's freezer. Cold storage space is limited and expensive.
- **Compliment good-quality work or extra effort.** Tell any deserving person involved in the work at the butcher shop that you appreciate their commitment.
- **Pay your bill promptly.** You don't like to wait on your payment from customers, so don't make your butcher wait either.

Exploring Your Different Selling Options

When evaluating how and where to sell your products, take the time to find the best situation for you — the one that optimizes your time and skills and reaches the most potential customers. The following sections show you some of the most common options.

Selling direct off the farm

Selling direct off the farm can be as simple as hanging a sign on the pasture fence that says "Freezer Beef for Sale" and listing your phone number and e-mail address. With this approach, you don't need any equipment or extra personnel. You can just make arrangements to have customers pick up their meat at the processing plant.

As your beef business grows, you may consider having a physical store on your farm. Just be sure you comply with all zoning and building codes before you get started on the project.

Unless your farm is convenient and close to a major population center, it may be hard to justify staffing a store full-time. Consider limited hours like all day Saturday and one evening a week. You can also think about partnering with other food producers to expand the food selection offered so customers can get their beef, eggs, honey, and produce all at one spot.

Participating in farmers' markets

Farmers' markets are great venues for selling farm-raised food because they bring interested, eager customers right to you for a specific and concentrated amount of time every week. If you can find an established market looking for a beef vendor, you could have an awesome opportunity on your hands.

The challenge is that many of the popular and successful farmers' markets already have established meat vendors, so new sellers are often put on a waiting list. But don't give up hope. Vendor turnover can be relatively quick, and you may get a spot at the market sooner rather than later.

New farmers' markets are continually popping up, and if you're just beginning in the beef business, these newer markets may be a more relaxed way of getting started. Being part of a growing market gives you the chance to learn and expand your business gradually with less pressure and commitment than trying to supply and conquer three or four busy markets every week.

Participating in farmers' markets is time consuming and requires plenty of hard work. You need to pick up your meat from the processor, pack the meat in coolers or freezers for transport, load all the other materials that are needed to set up a weatherproof mobile store, travel to and from the market, set up your stand, staff your booth for four or five hours, pack everything back up, put it all away when you get home, and then go take care of your cattle. So if your heart isn't in it, think twice about tackling this adventure, and instead choose one of the other options in this chapter.

Harnessing the power of the Internet

Because our increasingly mobile society is moving to more urban and suburban areas, people are losing connections with their rural food-producing contacts. As a result, more and more families who want to buy food direct from the farm are turning to the Internet to find the products they're looking for. You can tap into this growing market by having a website for your beef business.

Effective websites can be as simple as a one-page site listing your contact information and the products you have for sale. If you have the time and expertise (or are willing to learn), your website can be much more in-depth and include photos of your farm, cooking and recipe ideas, and even an online shopping cart and link to pay over the web.

There are lots of inexpensive web page templates and services that can get your farm on the web in a matter of hours. Take a look at our website (`www.royerfarmfresh.com`) for some ideas on how to get started on yours.

Working with restaurants

Providing beef to restaurants can be a viable selling opportunity. For some producers, it's the main outlet for their meat. Offering free samples is a good way to introduce yourself and your product to potential restaurant customers. If you can establish good working relationships with a few chefs, it can mean having consistent buyers for a large volume of meat. When your beef is served in restaurants, it brings increased creditability and recognition to your farm.

Selling beef to restaurants has some drawbacks:

- Often chefs expect the meat to be fresh, not frozen, which dramatically shortens the time you have to get your beef sold.
- You'll probably need several different accounts that use a variety of cuts so you can keep your inventory balanced.
- The restaurant may expect you to supply beef without giving you much lead time to get the product cut, wrapped, and delivered.
- You may be asked to give terms, such as providing the beef now, but getting paid at a later date.

Promoting and Marketing Your Beef

Suppose you've been working hard to raise wholesome and nutritious meat for your family, but now you're ready to share it with the world (or at least with a few willing friends, neighbors, and customers). This isn't the time to be modest. Instead, be proactive about telling potential customers what you have to offer. To successfully promote your beef, you want to communicate correctly and with enthusiasm both in person and through advertising. We explain everything you need to know in the following sections.

Beefing up on cattle-selling terminology

If you spend any time around the cattle industry, you'll hear numerous buzzwords describing how cattle are raised and how cuts of meat are processed. Don't ignore this lingo. You want to be clued in on this terminology so you can be clear in your promotional materials and during your interactions with customers about how you raise and feed your cattle and process your beef. Here are a few terms you should know:

- **Dry-aging:** *Dry-aging* is the process of letting a beef carcass hang in a refrigerated cooler for several days (usually 7 to 21) so naturally occurring substances in the beef can break down proteins and connective tissues. This process makes the beef more tender and palatable. You need good air flow for proper dry-aging.

  When you dry-age beef, water evaporates from the carcass, resulting in fewer pounds of saleable product. Dry-aged beef also has a more intense beef flavor and less watery cooking waste than wet-aged beef.
- **Grass-fed:** No official definition of the term *grass-fed* exists. This lack of clarification is in part due to the fact that almost all beef cattle eat grass/

forages at some point in their life and most beef comes from grass-fed animals. If your cattle are *100% grass-fed* and have only consumed pasture and harvested forages for their entire lives, refer to them as such. If they have been fed a diet of pasture and corn, refer to them as *grass/grain fed.*

- **Organic:** According to the USDA, *organic* is a labeling term indicating that the food has been produced through approved methods that integrate cultural, biological, and mechanical practices that foster cycling of resources, promote ecological balance, and conserve biodiversity. Third-party agents certify that farm operations comply with the USDA organic standards. To find out more about what's involved in organic production, head to `www.usa.gov` and search for "Title 7 National Organic Program." The first result should link you to the federal provisions.
- **Pasture-raised:** You can label your cattle as being *pasture-raised* if they never live in a confinement feedlot (a dirt lot) situation. Some consumers are interested in pasture-raised meats because the animal manure is spread out over more space and not concentrated in one area so there is less likelihood of pollution. They also think that a pasture is a nicer living space for animals than a feedlot.
- **Quality grade:** The USDA uses a *quality grade* to differentiate the expected differences in the eating quality among beef carcasses. USDA employees assess and provide eight USDA quality grades for beef: Prime, Choice, Select, Standard, Commercial, Utility, Cutter, and Canner. Eating quality is usually best for beef graded prime and decreases in the order listed. Quality grades are based on the animal's maturity and marbling (intramuscular fat) as well as the firmness, color, and texture of the beef.
- **Wet-aging:** The *wet-aging* process involves refrigerating cuts of beef in a vacuum-packed plastic bag for a few days or longer. Just as in dry-aging, wet-aging allows the enzymes in the beef to break down some of the proteins and connective tissue, thereby increasing tenderness. However, without any evaporation of liquid from the meat, the flavor won't be concentrated as with dry-aged beef.
- **Yield grade:** The yield grade is an assessment that estimates the number of boneless, closely trimmed retail cuts from the round, loin, rib, and chuck of a beef carcass. USDA yield grades are 1, 2, 3, 4, and 5; Yield Grade 1 is the highest-yielding carcass, and Yield Grade 5 is the lowest. Yield grades are determined by USDA employees.

Spreading the word about your products

Even if you have the best-tasting, lowest-priced beef in the world, no one will buy it if they don't know it exists! To get your beef business off to a successful start, you need to enthusiastically and continually promote your product. Some ways you can spread the word include

- **The shirt on your back:** A shirt or jacket with your farm logo or advertising slogan can turn you into a walking billboard.
- **Business cards:** Always have some cards handy, because you never know when you may meet your next potential customer.
- **Farm tours:** Schools are often looking for low-cost, educational field trips. While the kids learn about farm life, you can educate the adult chaperones about the benefits of your beef.

  Check with your insurance agent to make sure you have the proper coverage if you're going to conduct farm tours.

- **Civic organizations:** Home economics clubs, volunteer organizations, and business groups often have local speakers in conjunction with their meetings. These are great opportunities to tell your farm's story to many potential customers.

While it's great to be your own spokesperson, you can't be out on the promotional trail 24/7. After all, who would take care of the cattle? As you look to grow your business, take the time to develop a marketing plan. To make the most of your budget, determine who you're trying to reach and stay focused on those potential customers. Here are some ideas:

- **Create a website.** A website is a great, low-cost way to get in front of many interested customers. Get listed on or linked to national and regional websites that promote small-scale farming or beef production. Numerous sites are specifically designed to connect producers with consumers, and most of them are free.

  Be sure your website contains lots of key words people may use while doing an Internet search, including "beef," "local," "farm-raised," "grass-fed," "family farm," "freezer meat," "organic," and "natural."

- **Advertise at events.** Chili cook-offs, wine festivals, or Earth Day celebrations where your potential customers may gather can be a good venue for distributing handbills and flyers or setting up a display.
- **Submit press releases.** Don't be afraid to toot your own horn. If your farm receives recognition for environmental practices or donations to a food pantry, let the local press know. Many newspapers and even TV stations are looking for good human/food/animal interest stories.
- **Design printed materials.** Create and use print ads, direct mail, billboards, farm signs, business cards, and brochures. No matter what printed material you use, keep the look consistent. Use the same colors, fonts, and tag lines so you and your brand become easily recognizable to potential customers. Also make sure that your website matches your printed material.

Part V
The Part of Tens

In this part . . .

In this part are two quick chapters. The first one looks at ten pitfalls to avoid when raising cattle, and the other shows interesting bovine behaviors that you need to watch out for.

Chapter 16

Ten Common Mistakes to Avoid When Raising Cattle

In This Chapter

- Planning ahead to avert problems
- Becoming familiar with missteps you can make

We have more than 55 years of combined cattle-raising experience, and over the years we've made most of the mistakes we talk about in this chapter. Such "teachable moments" aren't fun things to experience, but, as we can attest, they are survivable. Although experience may be the best teacher, simply reading and remembering is a lot less work and stress. So here are ten mistakes to avoid when raising beef cattle!

Tolerating Weak Perimeter Fences

If you could see our mental psyche, you would see the scars left from raising cattle in pastures with bad perimeter fences. At one point, we got daily calls during August and September when the pastures were bare that our cattle had escaped for the lush green corn across the saggy, rusty barbed wire that tried to keep them contained.

Quality fencing is one of your most important assets when raising cattle. It's particularly important if your pasture is less than ideal. Cows are no fools. When their pasture is no longer top quality (or nonexistent), they will search for something better. It takes a substantial fence to stop a determined 1,200-pound bovine, so if you have weak spots in your fencing, your animals are going to get out. And we bet you'll feel pretty sheepish when your neighbors pay you daily calls about your roaming cattle.

The passage of time lets us chuckle at some of the adventures brought about by poor fences, but in all seriousness, you must remember that you put motorists and your cattle at risk when you have weak fences. Chapter 4 gets into the details of what you need for a strong and safe fence system.

Before you buy any cattle, be sure to get your fences in good shape; it's an investment that pays for itself every day your phone doesn't ring with someone on the other end saying, "Your cows are out!"

Running Out of Grass

When it comes to pasture, it's always better to have too much than not enough. Unfortunately, pastures don't grow at a consistent rate 12 months of the year. During late spring and into June, the grass grows like crazy, and you'll probably have extra forage. But if you build up your cattle numbers to match this time of peak production, you'll probably be short of forage the other ten months of the year.

Your best bet is to match the number of cows to the amount of winter feed your pastures produce. During the periods of high growth, you'll likely have extra, but you have some options for taking advantage of a forage bonanza. Consider the following:

- **Match your livestock needs to your pasture production.** Our cows give birth on pasture in May just as the grass is really taking off. As we discuss in Chapter 5, the nutritional needs of the lactating mother are greatest in the weeks after calving, so it works well to have hungry cows benefiting from the abundant grass.
- **Raise extra stocker animals.** If you live reasonably close to areas where cattle are raised as stockers over the winter, you may be able to continue feeding these cattle on your excess pasture. You can purchase the animals or work out an arrangement to be paid a set amount for each pound the cattle gain. See Chapter 13 for more ideas on growing stocker calves.
- **Make hay.** Cattle can harvest pasture more cheaply than people can cut and bale hay (plus they fertilize as they go). So you're better off making and feeding as little hay as possible. However, if your pastures are becoming mature and going to waste, you may want to consider putting up hay. You can hire professional custom hay balers to harvest your crop for a set fee or a share of the hay.

Forgoing a Calving Ease Bull

A *calving ease bull* is a bull that's genetically more likely to produce calves that are born easily compared to the average bull. Calving ease is usually due to a smaller size at birth, but it can also be related to the shape of the calf's body.

If you're using artificial insemination to breed your heifers and cows, use semen from a bull with a proven record of calving ease. If a bull breeds your cows naturally, make sure his calving ease expected progeny difference (EPD) is at least better than average. (Refer to Chapter 7 for more on EPDs.)

Some producers are concerned that calves sired by a calving ease bull grow slowly and don't get as big as other calves. This belief may have been true in the past. But now that EPDs have been widely used for several decades, you can find plenty of bulls that produce calves that are born easily and then take off growing fast and big.

An unassisted birth means less work for you and a greater likelihood of a strong, vigorous, and profitable calf. An unassisted birth usually means the cow has to work less, so she has more energy to get up and take care of her calf. She's also less likely to suffer injury during an easy labor as opposed to a difficult delivery. The calf goes through less trauma during an easy delivery, thus allowing it to get up and eat more quickly. The animal is also less likely to get harmful fluid in its lungs.

Neglecting to Keep an Adequate Backup Water Supply

Water is vital to the health and well-being of your cattle. They need water for adequate growth, reproduction, and lactation. During the heat of summer, a cow with a nursing calf drinks nearly 20 gallons of water a day. (Head to Chapter 5 for more on the routine water supply needs of your cattle.)

Having a backup water supply plan in place is important in case you have an emergency. Here are some scenarios you want to be prepared to deal with:

- **Electrical outage or service interruption:** If you depend on electricity to run your pump, consider what you'll do during a long-term power outage. Similarly, if your water is supplied by a municipality, what will you do if service is disrupted? If you do have an electrical outage or service interruption, it helps to have ground water access, such as streams or ponds, for your cattle. You also can use a large container tank to haul water to your cattle from another source.

- **Broken pump:** Consider what happens if your pump needs repair or replacement. Without a pump, your cattle will need water from another source. On our farm, we have four different pumps so that if one pump goes out, we can always move the cattle to a pasture serviced by a different pump. We've also connected the water lines for two different pumps so that one pump can do the work of two in case of a breakdown.
- **Drought:** If your cattle depend on streams or ponds for their water supply, think about what you'll do if the stream or pond dries up. You can haul water as a stopgap measure, but you may want to drill a well for an easier and more consistent supply of water.

Lacking a Marketing Plan

Raising cattle is fun and rewarding, but being profitable makes it even better! If you start out your herd willy-nilly without thinking about your goals and how you plan to market your cattle to reach those goals, your wallet may end up a little lighter than you may expect.

The nice thing about cattle is you can almost always find a buyer regardless of the type of cattle. However, to command top dollar for your livestock, you need to raise what the buyers want and be confident and positive when promoting your cattle.

A successful marketing plan should tie in with your cattle-raising goals. Before you buy your first cattle or expand your herd, plan how you'll sell those new animals or merchandise their offspring. If you want to raise beef to sell to others, your marketing plan should focus on how, where, and when you'll sell meat. If you want to have stocker calves, you need to research what weight, color, and type of calves you need to produce.

Buying Bargain Cattle

The old saying, "You get what you pay for" is almost always true with regard to cattle. Cattle that are inexpensive to purchase initially are often more expensive in the long run. Discounted cattle may be sick or have bad dispositions. They also can be old and nearing the end of their productive lives. Avoid buying old cows unless they're priced very cheaply. They may not have fully functioning udders, so they may not be able to raise a calf. They may also be slow to rebreed (if they can become pregnant at all).

You can't pay us to take sick cattle or animals with bad dispositions. We have worked too long and hard to build up a healthy, well-behaved herd to have it ruined by a few bad animals. It only takes one contagious animal to spread disease to an entire herd. For the most part, cattle are hardy, but plenty of mean germs can make cattle feel miserable, cause the loss of a pregnancy, or even result in death.

Just as disease can be spread, so can bad attitudes. One temperamental animal can negatively impact the manners of the entire herd. Temperamental animals are more likely to escape from their pasture and are harder to corral when you need to handle them. Wild cattle don't grow as quickly and have a tendency to produce tougher meat.

To make sure you don't get stuck with bargain cattle, you need to know the typical prices. Numerous third-party price reports are available for all classes of cattle. Weekly prices are listed in farm newspapers, or you can search the Internet for the cost of commodity cattle. For purebred cattle prices, visit breed association websites for information or check out cattle magazines for sale reports.

Purchasing Cattle for Looks and Not Function

An animal with a glossy coat, bright and clear eyes, and a muscular and well-proportioned body is most often a healthy and high-performing individual. It's a pleasure to see these animals. However, avoid getting too wrapped up in the looks of your cattle. You may end up overlooking these other important traits:

- **Cost:** If you find your cattle need to be provided with more grain and more hay every year to maintain their looks, you may need to reevaluate their value to your herd.
- **Reproductive ability:** Keep in mind that your breeding stock should be able to consistently reproduce. A pretty cow becomes an expensive pasture decoration if she doesn't have a calf every year. And even if a bull has won many champion ribbons at the fair, he's not doing his most important job if he's not siring calves.

Expected progeny differences (EPDs) are a valuable, objective selection tool that can help you choose animals that will function in the areas of calving ease, growth, milk production, and carcass merit. They're an excellent complement to the more subjective visual appraisal of cattle anatomy.

Skimping on Training before the Cattle Show

Exhibiting cattle at shows or sales is expensive and takes a lot of time (see Chapter 14 for details). Make the most of your efforts by fully training your cattle beforehand. Your cattle need to calmly stand while you position their feet, and they need to move obediently when you lead them. To maximize their show potential, try to expose them to as many different experiences as possible before the show so they quickly adjust to new situations.

Here are some ways that you can get your cattle ready for the big day:

- **Set up practice shows all around your farm.** In other words, don't always practice in the same pen. Work with your cattle away from other cattle, with unfamiliar cattle, with a loud radio playing in the background, or with a piece of garbage blowing by. You want them to get used to everything that may distract them.
- **Introduce your cattle to new people.** You especially want cattle to get used to being around people with loud voices. Also expose your cattle to people of different sizes. If cattle are only used to being around adults, children can really startle them.
- **Get the cattle accustomed to eating and drinking out of the same feed and water equipment that they'll use at the show.** You want the cattle to be comfortable with their feed and water gear so they will keep eating and drinking while away from home and look their best for the show.
- **Be realistic.** Even with a lot of practice at home, cattle usually don't behave perfectly at their first show. If feasible, do a practice run at a small, local event before going to the "big" show. The smaller show will be a good dress rehearsal for everyone.

Failing to Set Proper Customer Expectations

Repeat after us: Satisfied customers are critical to the success of all businesses. It takes a lot less effort to keep a current customer than it does to find a new one. Plus, word-of-mouth advertising from satisfied customers is free and effective. To cultivate happy, returning customers, be clear and direct about the benefits and drawbacks of your product — whether that product is a show calf, a herd bull, or a package of steaks.

When selling cattle, be sure to put important information about the animal in writing so you have no squabbles over expectations later. Consider noting the following so the buyer is crystal clear on what she'll get:

- Provide the customer information about the sire, dam, birth date, growth data, and immunization schedule of the cattle she's thinking about buying.
- Make it clear that although your cattle are calm, they aren't tame, well-trained show cattle (unless they are; and in that case, play it up!).
- Talk about what the cattle are accustomed to eating so the buyer can decide whether your cattle can adapt to her specific feeding program.

Also make a concentrated effort to communicate clearly with your beef customers. Again, put important facts — price, quantity to be received, and delivery date — in writing. Before a customer buys any beef, talk about what she can expect, including the state of the meat (fresh or frozen), its fat content, and a description of the flavor. You can encourage customers to buy a package or two to try before they make bulk purchases.

Making Raising Cattle Complicated

When starting a new endeavor like raising cattle, you can be overwhelmed with all there is to learn and do. Don't get hung up on every little detail; you'll become too paralyzed by information overload to ever take action. Raising cattle doesn't have to be difficult or complex, just stay focused on the main, basic keys to success.

To have a positive experience raising cattle, all you really need to do is

- Decide why you want to raise cattle
- Buy healthy animals that fit with your goals
- Provide a clean, spacious living environment with good fences
- Supply wholesome feed and plenty of water
- Handle your cattle calmly
- Practice preventive health maintenance
- Develop and implement a marketing plan
- Ask for help if you or your cattle need it

Chapter 17

Ten (or so) Bizarre Bovine Behaviors . . . and What They Mean

In This Chapter

- Deciphering cattle behavior
- Communicating with your cattle

Cattle will tell you what they require to be healthy and content. You just need to be able to understand what they're saying. Being familiar with the ways cattle communicate and how they react to their environment helps you develop a positive relationship with them. In this chapter, we provide a list of bizarre cattle behaviors that help you determine exactly what your animals are telling you.

Eating Old, Brown Grass When Fresh Forage Is Available

If you have cold, drab winters like we do in the Midwest, the arrival of spring is always welcome. So it only makes sense that when the cattle are first turned onto the new grass they hardly make it through the gate before they begin gorging on the fresh, green sprouts. However, you may instead notice that over the next day or two they ignore the beautiful tender forages and instead chow down on dried up grass, leaves, and stems. When they do this, the cattle are telling you that something is missing from their early spring pasture diet.

Young, spring forage growth is often *washy.* This term describes pasture with a high water content and lower nutrient content. In their highly vegetative state, these forages are often low in fiber. To keep the rumen functioning properly, the cattle begin craving more mature plant material that can provide fiber.

To keep your cattle from digestive upset and from mashing down fences to get to the old pasture on the other side, provide free-choice medium-quality grass hay to the cattle while they adjust to their new diet and while the pasture matures. A properly balanced diet also helps reduce the next fun behavior . . .

Squirting Projectile Manure

The manure from healthy cattle can range in color from yellow to green to brown depending on their diet. Its consistency should be about like oatmeal. If the manure from your mature cattle is bright green and comes squirting out and hits the ground 3 or 4 feet behind the animals, you need to make some adjustments to their diet.

This type of projectile manure is often seen in cattle grazing new, lush spring pastures and happens because the animals don't have enough fiber in their diet. Without adequate fiber, the digestive process occurs too quickly, preventing water — and nutrients — from being properly absorbed by the body. Make sure your cattle have access to medium-quality grass hay to meet their fiber requirements and to protect yourself from an unexpected manure encounter.

Not Drinking from the Water Trough

Like humans, cattle need water to maintain their health. If you notice that your cattle aren't drinking normally from their water troughs, carefully inspect the area to be sure nothing is deterring them from enjoying a refreshing drink. (For specifics on cattle water requirements, check out Chapter 5.) Be on the lookout for:

- **An electrical short:** Contact between the electric fence and water trough can make for a shocking drink. So can problems in the electric motor or heater of an automatic water trough.
- **An unwelcome visitor in the trough:** Sometimes small animals such as squirrels and birds stop by the trough for a drink and accidentally drown. Their remains can soil the water and discourage your cattle from drinking. Also be on the lookout that your cattle haven't confused the water trough for a toilet!

- **A change in the taste of water:** If you have municipal water and the lines are being flushed for maintenance or repairs, the water may have an unpleasant taste or smell causing the cattle to refuse to drink it. Usually, the water gets back to normal — within 24 hours or so — before the cattle are at risk of health problems. However, if the weather is very hot and humid, supply your cattle with water from another source so they don't overheat.

Galloping around the Pasture

All healthy cattle — even mature mother cows — sometimes feel like running. Even if you strive to maintain a low-stress, quiet environment for the livestock, sometimes the animals just need to let off some bovine steam!

Don't be concerned if your cattle run around every few weeks or so. However, if you see your cattle running during warm summer and fall days on a more frequent basis without any visible reason, your cattle may be under attack from heel flies. *Heel flies* are flying insects that look and sound like honeybees. Their buzzing sound makes cattle highly agitated and often results in panicked running across the pasture. (See Chapters 9 and 10 for details on dealing with heel flies.)

It's perfectly normal for all cattle to feel a bit exuberant at times and to take off bucking and galloping, but you should use extreme caution around cattle that are acting this way because the standard cattle rules about flight zones and predator-prey relationships (discussed in Chapter 8) do *not* apply. We have seen cattle that are so excited they chase people and come close to charging them, even though the animals are in an extremely good mood and don't feel at all threatened.

If you think there's any chance your cattle may be feeling frisky, don't enter their pasture on foot. Take a vehicle, a horse (if your cattle are familiar with them), or an ATV. These modes of transportation offer you protection from a bouncing, running cow.

If your cows catch you by surprise and start acting lively when you're on foot, it's time for you to ditch the quiet manners and take charge of the situation:

- Speak loudly and firmly to the cows to get their attention with authority.
- Wave your arms slowly or shake a stick to make yourself appear larger and in charge.
- Maintain eye contact with the herd as you slowly back away.

Pacing along the Fence

If your cattle are pacing back and forth along the fence, they're agitated about something. You need to find out what's causing their distress so you can try to remedy the situation. For example, check out these common instances of pacing and how you can address the issue:

- **A bull may be restless when he wants to get in with nearby cows.** If possible, move him away from the temptation so he doesn't crush your fence.
- **Cows and calves pace at weaning time.** Take advantage of this behavior by placing feed bunks and water troughs perpendicular to the fence so the young calves easily find feed and water.
- **Cattle may pace if their food or water supply is spoiled and they're hungry or thirsty.** A wild animal may have gotten into the free-choice feed or water trough and spoiled it. Simply refresh the trough so the cattle can have a bite or a drink.
- **Cattle that you've moved to a new pasture may prowl around the perimeter of the fence.** Cattle will almost always want to check out the boundaries of their new living area. To give your cattle time to calm down after a ride in the livestock trailer, have them rest overnight in a secure pen or corral before releasing them into their pasture.

When you notice pacing of any kind, try to defuse the situation. Stay nearby and watch the herd until they settle down. In an excited state, they may push too hard on a gate or low spot in the fence and escape.

Balking at Shadows or Holes

Cattle have amazing peripheral vision because their eyes are on either side of their head. However, this anatomy also means that their depth perception is pitiful. A shadow or small hole can look like a deep, dark abyss to a cow.

If your cattle are balking at going down an alley or loading in or out of a chute or trailer, check for shadows. An animal frightened by shadows may abruptly stop walking and refuse to move. Or it may jump out of the chute instead of walking out.

Try to look at things from your cattle's perspective and work to eliminate patterns of light and dark. Try the following:

- Use bright lights directly overhead alleys and handling facilities to reduce the chance of light angling through fences and gates and making shadows on the ground.
- When loading cattle from the barn into a livestock trailer or unloading them into a barn, back the trailer into the barn to eliminate a drastic change in light in the gap between the barn and trailer.
- If possible when designing your handling facility, position the chute so it doesn't face directly into the sun. Cattle have a hard time seeing when light is shining directly in their faces, and they are hesitant to move under such conditions.

Bawling Frantically

Even after many years of raising cattle, we're still impressed with the patience exhibited by young calves. A mother will settle her calf into a protected spot in the pasture and leave it while she goes to graze. The calf lies still for hours until its mother returns — unless, of course, it gets scared. Then it jumps up and takes off running and bawling frantically.

If you see one of your calves searching for its mother, give the calf several minutes to find her. Any cow with decent maternal instinct will hear her calf's call and hurry to join it. If the pair doesn't reunite after 10 or 15 minutes, investigate the situation to see whether you need to help the reunion process along.

Rub-a-Dub-Dubbing on Anything and Everything

Cattle like to rub . . . a lot! Rubbing is the best way for them to scratch all their bovine itches. It's also the way cattle help speed along the process of shedding their old winter coats when warm spring temperatures arrive. You don't need to be concerned with rubbing unless the cattle develop bald or raw spots on their skin. These signs may indicate lice or mange. (See Chapter 10 for more on these conditions.)

If your cattle, especially bulls, tear up fences or equipment with their rubbing, you may need to install electric wires to prevent damage. Unless barbed wire is tightly strung and very sharp, it may actually serve as a scratching spot instead of a deterrent.

I like it when you scratch my . . .

One good way to bond with your tame cattle is to scratch them, especially in places on their body that are hard for them to reach, including the following:

- **Chin:** They like being rubbed under the chin.
- **Brisket:** With a show stick, you can scratch the area between the front legs.
- **Lower belly:** Use a show stick to scratch from the front legs to the navel.
- **Tailhead:** A favorite scratching spot for some bovines is right at the base of the tailhead. But, of course, you only want to scratch here if they aren't too dirty.

Warning: Never rub an animal on the top of its head in the poll area. This interaction makes them more likely to try and butt you in the future.

Curling Their Lips

Don't worry, your steer isn't snarling at you when his lips are curled. Several different animal species, including cattle, exhibit the *flehmen response,* which is an upward curling of the top lip. This positioning of the lip helps an animal better detect and analyze smells.

For instance, bulls have a flehmen response when they're smelling the urine of a female, especially when she's in heat. Both genders curl their upper lips when they encounter new feedstuffs or when new cattle join the herd. They also use the flehmen response to help detect predators.

Index

• Numerics •

100 percent grass-fed beef
  customers for, 272–273
  feedsfuffs for, supplying, 273
  overview, 271–272
  as selling terminology, 309
  taste of, 272
  type of cattle suited for, 273–274

• A •

abomasum, 28
abortions, 237–238
accommodations, 12–14
acidosis, 102
active labor, 233–236
ADG (average daily gain), 40
advantages of raising beef cattle, 9–11
AI (artificial insemination)
  advantages of, 226
  disadvantages of, 226
  overview, 225
  requirements for, 227
  steps for, 227–229
Alarm moo, 153
alfalfa, 109
American Brahman cattle, 50
amino acids, 80
anatomy of cattle, 25–31
Angus cattle, 43–44
animal husbandry
  ear tags, using, 187–189
  first-aid kit, stocking a, 183–185
  injections, administering, 185–187
  wounds, treating, 189
anthelmintics, 209
antibiotic therapies for respiratory diseases, 193
appearance, purchasing cattle based only on, 317
appetite, monitoring, 166
arrival of cattle at your farm, minimizing stress during, 143–144
automatic waterers, 106
average daily gain (ADG), 40

• B •

B vitamins, 91
balking at shadows or holes, 324–325
banding, 257
barbed wire fencing, 59
bargain cattle, buying, 316–317
barn blind, 134
Barnes dehorning, 257
basic requirements
  accommodations, 12–14
  nutritional requirements, 12
  water source, 12
bawling frantically, 325
BCS (body condition scores), 214–216
bedding materials, 75–76
beef congresses, 278
beef farms as source for bottle calves, 128
Beefmaster cattle, 51
behavior of cattle. *See* cattle behaviors
Belted Galloway cattle, 51
Bermudagrass, 111
big projects, planning for, 14–15
birthing positions, 234–236
birthing spaces, setting up, 230–231
birthing time, 163
bizarre cattle behaviors. *See* cattle behaviors
black cattle, providing shade and water for, 37
blind spot, 154
bloat, 211–212
body condition scores (BCS), 214–216
body language of cattle, 151–152
body temperature
  checking, 168–170
  raising, 251
Bos indicus cattle, 43

Bos taurus cattle, 43
bottle calves
beef farms as source for, 128
caring for, 129–130
dairy farms as source for, 128
feedlots as source for, 129
finding, 128–129
overview, 128
purchase prices, 15
sale barns as source for, 129
selecting, 128–130
bottle feeding, 247–249
bovine respiratory syncytial virus (BRSV), 195
bovine virus diarrhea (BVD), 194–195, 220
Brangus cattle, 51
breaking halter, 281
breathing difficulties, helping newborns with, 242–243
bred cow, 24, 131–132
bred heifer, 24, 131–132
breech position, 236
breed associations, junior shows sponsored by, 278–279
breeding season, 176–177
breeding soundness exam, 223–224
breeding stock
defined, 24
handling
bull, evaluating attitude of, 161–162
calves, working with, 163
equipment for, 161
new mothers, working with, 163
overview, 160–161
breeding up, 47
breeds. *See also specific breeds*
British, 43–46
cattle hair, deciding on breed by, 37
characteristics of, 39–40
choosing, 35–42
composite, 50–51
Continental, 46–50
crossbred cattle, 40, 41–42
heritage, 51
horns, choosing a breed based on, 38–39
nearby sources for, deciding on breed by, 36–37
overview, 35, 42
purebred cattle, 40–41
brisket, 26
British breeds
Angus cattle, 43–44
Hereford/Polled Hereford cattle, 44–45
overview, 43
Red Angus cattle, 45–46
Shorthorn cattle, 46
BRSV (bovine respiratory syncytial virus), 195
brucellosis, 199–200, 219
brushing, 286–287
building insurance, 21
bull calf, 24
bulls
breeding soundness exam, 223–224
caring for, 224–225
castrating, 257–258
defined, 24
feeding program for, 99–100
leasing, 229–230
protein recommended for, amount of, 82
selecting, 221–223
bunk feeder, 102–104
bunk-broke, 130
business permits, 295
butchers
animal welfare, evaluating, 304
facilities and personnel, evaluating, 303
overview, 302–303
packages and labels, options for, 304–305
relationship with your, 305–306
buying cattle
conformation, judging, 139–140
from consignment sales, 134–135
contagious diseases, avoiding, 138
customer service, evaluating, 141–142
disposition, assessing, 138–139
from farms, 133–134
genetic diseases, avoiding, 138
healthy bovines, signs of, 137
from Internet, 136
overview, 132–133

performance measurements, 139–140
from sale barns, 135–136
from test stations, 135
unhealthy bovines, signs of, 137
BVD (bovine virus diarrhea), 194–195, 220

• C •

calcium, 85, 88
calves
bawling frantically, 325
castrating young bulls, 257–258
creep feeding, 254–255
defined, 24
ear tags used for identifying, 244
horns, removing, 256–257
identifying, 243–245
tattoos used for identifying, 244–245
vaccination schedule, 255
weaning, 258–260
calving ability, 43
calving ease bull, 315
calving pasture, creating a, 230–231
calving season, 175–176
calving shelter, building a, 231
carbohydrates, 83–84
carcass, 39
carcass merit, 39
castration, 150, 257–258
cattle behaviors
balking at shadows or holes, 324–325
bawling frantically, 325
curling lips, 326
fence, pacing along, 324
manure, squirting projectile, 322
monitoring for health of cattle, 166–167
old grass, ignoring fresh green grass for, 321–322
pasture, galloping around, 323
rubbing, 325
understanding, 148–150
water trough, not drinking from, 322–323
cattle shows. *See* exhibitions
cervix, 31
characteristics of breeds, 39–40
chargers for electric fencing, 62–63
Charolais cattle, 47
chemical dehorning, 256
Chianina cattle, 47–48
clamping, 257–258
cleaning shelters, 71
cobalt, 86, 88
cold carcass weight, 301
cold weather, steps for dealing with, 182–183
colostrum, 96, 247–248
commodity pricing, 301
complicated, avoiding the mistake of making raising cattle, 319
composite breeds, 50–51
composting, 116–118
concentrates, 94–95
conformation, judging, 139–140
consignment sales, 134–135, 290
contagious diseases, avoiding, 138
Continental breeds
Charolais cattle, 47
Chianina cattle, 47–48
Gelbvieh cattle, 48
Limousin cattle, 48
Maine-Anjou cattle, 48
overview, 46–47
Saler cattle, 48–49
Simmental cattle, 49
Texas Longhorn cattle, 50
cool-season forages, 123, 124
Cooperative Extension Service, 73
copper, 86, 88
core sampler, 94
corn cobs as bedding materials, 76
costs plus markup, 301
county fairs, 278
cows
defined, 23, 24
protein recommended for, amount of, 82
creep feeding, 97, 254–255
critical temperature point, 182–183
crossbred cattle, 40, 41–42
crowding pen, 66
crude protein percentage, 81
cud, 27
cud chewing, 27, 166

cull, 24
curling lips, 326
custom grazing
  contract for services, creating a, 270–271
  finding cattle/owners, 270
  overview, 268–269
  pasture resources needing to match with number of cows, 269–270
customer expectations, failing to set proper, 318–319
customer service when buying cattle, evaluating, 141–142
custom-exempt processing facilities, 294
cuts of meat, 299

• D •

daily chore commitments, 14
dairy cattle, 32
dairy farms as source for bottle calves, 128
dam, 24
dehorned cattle, 38
dehorning calves, 256–257
delivery
  active labor, 233–236
  birthing positions, 234–236
  birthing spaces, setting up, 230–231
  breech position, 236
  calving pasture, creating a, 230–231
  calving shelter, building a, 231
  dystocia, 234
  monitoring stages of labor, 233–237
  paralysis, temporary, 239–240
  postpartum, 237
  pre-delivery, 233
  supplies for, 231–232
  of twins, 237
Devon cattle, 51
dewclaw, 26
diptheria, 192
direct off farm as selling options, 306
direct threat position, 162
disposition, assessing, 138–139
double-span end assembly, 58
drainage for shelters, 70
dry cow, feeding program for, 98
dry forages, 94
dry matter basis, 81, 119
dry matter production, 120
dry-aging, 308
due date, determining, 220–221
dystocia, 234

• E •

ear tags, 187–189, 244
ears, reading body language of, 151
elastration, 257
electric fencing
  chargers for, 62–63
  overview, 62–63
  safety issues, 63–64
  setting up and maintaining, 64–65
electric twine, 60
emergencies, planning for, 15
emergency healthcare, 16
end assemblies, 58
endophyte, 109
English breeds. *See* British breeds
environmental adaptability, 40
EPDs (expected progeny differences), 140–141, 218, 317
esophageal feeder, 247–248
exhibitions
  breed associations, junior shows sponsored by, 278–279
  county fairs, 278
  expos or beef congresses, 278
  goals for, 276–277
  grooming your animals for, 285–287
  jackpot or preview shows, 277–278
  junior shows, 277
  major livestock exhibitions, 279
  open shows, 277
  presenting your animals and yourself in the best possible way, 285
  show ring protocol, following, 287–288
  show stick, using a, 280
  state fairs, 279
  supplies for, 283–284
  training your animal for, 280–283, 318
  types of, 277–279

existing structures, repurposing, 72–73
expos or beef congresses, 278
extension agents, 18–19
external (permanent) fences
  barbed wire fencing, 59
  overview, 56
  woven wire fencing, 56–59
external parasites
  flies, 204–206
  grubs, 206
  lice, 207
  mites, 208
  overview, 180–181, 204
  ringworm, 207–208

• F •

F1 crossbred, 41
F2 generation, 41
facilities. *See* handling facilities
family project, raising beef cattle as a, 11
farm liability insurance, 21
farm personal property insurance, 21
farm policy, 296
farm ponds, 125–126
farmers' markets as selling option, 306–307
farms, buying cattle from, 133–134
fat cattle (fats), 24, 84
feces, monitoring, 167–168
federally inspected meat processing plants, 294
feed dealer, 19
feed expenses, calculating, 15–16
feed tag, 80
feeder cattle, 24, 97
feeding program. *See also* nutritional needs
  for bulls, 99–100
  choosing, factors in, 92–93
  concentrates, 94–95
  customizing your, 90
  for dry cow, 98
  for feeder calves, 97
  finishing diet, 98
  forages, 93–94
  grain feeders, 102–104
  for growing calves, 97
  hay rings, 104
  introducing new feeds to diet, 101–102
  for newborns, 96
  nontraditional feedstuffs, 95
  for 100 percent grass-fed beef, 273
  for pregnant/lactating cows, 99
  schedule for feeding, 100–101
  for stocker calves, 97
  supplies, 102–106
  water sources, 104–106
feedlots as source for bottle calves, 129
feed-related problems
  bloat, 211–212
  founder, 211
  moldy feed, 210
  poisonous plants, 210
  sweet clover poisoning, 212
feed-through fly control, 206
female reproductive anatomy, 30–31
fence-line weaning, 259
fences
  external (permanent)
    barbed wire fencing, 59
    overview, 56
    woven wire fencing, 56–59
  internal (subdivision)
    electric fencing, 62–65
    gates, planning for, 60–62
    overview, 56, 60
    polywire conductor fences, 60
    semipermanent fences, 60
    types of, 60
  overview, 55–56
  pacing along, 324
  weak perimeter, tolerating, 313–314
  windbreak, 74
fertilizing soil, 114–115
fescue, 110
fetlock, 26
fighting among cattle, 149
financial commitments
  facilities and land, budgeting for, 17
  feed expenses, calculating, 15–16
  healthcare, budgeting for, 16
  purchase prices, 15

financial profit as benefit of raising beef cattle, 10–11
finishing, 273–274
finishing diet, 98
first-aid kit, stocking a, 183–185
first-calf heifer, 24, 132, 175, 217–218
flank, 26
flash grazing, 126
flehmen response, 326
fleshing ability, 44
flies, 180, 204–206
flight zone, 155–158
float valve, 105
Foghorn moo, 153
food safety permits, 295
forages
  characteristics of quality, 108–109
  grasses, 110–111
  legumes, 109–110
  overview, 93–94
  running out of, 314
  types of, 109–111
founder, 211
free-choice feeding, 101
freezer beef, providing, 297–298
frothy bloat, 211

• G •

gassy bloat, 211
gate
  moving cattle through, 157–158
  planning for, 60–62
Gelbvieh cattle, 48
genetic diseases, avoiding, 138
genetically true, 51
goiter, 86
grain feeders, 102–104
Grandin, Dr. Temple (expert in animal handling and welfare), 73
grass tetany, 85, 181
grasses, 110–111
grass-fed, 308–309. *See also* 100 percent grass-fed beef
grass/grain fed, 309
grazing
  custom
    contract for services, creating a, 270–271
    finding cattle/owners, 270
    overview, 268–269
    pasture resources needing to match with number of cows, 269–270
  flash, 126
  patch, 119
  rotational, 13, 121–122
  spot, 119
  spring and summer, 264
  winter, 264
grooming your animals for exhibitions, 285–287
ground-driven spreaders, 117
growing calves, feeding program for, 97
growing market cattle, protein recommended for, 82
grubs, 180, 206

• H •

H. somnus (haemophilus somnus), 196
hair, deciding on breed by, 37
halter-breaking, 281
halters, 281–282
handling cattle
  behavior of cattle, understanding, 148–150
  body language of cattle, 151–152
  breeding stock
    bull, evaluating attitude of, 161–162
    calves, working with, 163
    equipment for, 161
    new mothers, working with, 163
    overview, 160–161
  fighting among cattle, 149
  innate behavior of cattle, utilizing, 154–160
  moving cattle by using flight zone, 155–158
  noises made by cattle, understanding, 153–154
  preferences, recognizing cattle, 148

riding (mounting) cattle, 150
training cattle, 160
handling facilities
budgeting for, 17
of butcher, evaluating, 303
crowding pen, 66
designing, 66–69
ease of use of, 68
head gate, 66, 69–70, 161
holding chute, 66, 69–70
holding pen, 66
loading chute, 66
overview, 65
problems to look for and fix in, 159
safety issues, 68
selecting a site for, 65–66
space in, 68
for stocker calves, 268
strength of, 67
working chute, 66
working with cattle in a, 158–159
hauling cattle, 142–143
hay rings, 104
head gate, 66, 69–70, 161
heads, reading body language of, 151
health of cattle. *See also* herd health calendar; *specific diseases*
appetite, monitoring, 166
behavior, monitoring, 166–167
body temperature, checking, 168–170
budgeting for, 16
clostridial diseases, 196–198
cold weather, steps for dealing with, 182–183
early, recognizing illness, 165–168
emergency healthcare, 16
external parasites, 204–208
extreme weather, keeping cattle comfortable during, 182–183
feces, monitoring, 167–168
feed-related problems, 210–212
hot and humid weather, steps for dealing with, 183
internal parasites, 209–210
nutritional needs, seasonal, 181
parasites, controlling, 178–181
pinkeye, controlling, 181
pulse, measuring, 170
reproductive diseases, 199–201
respiration, observing, 171
respiratory diseases, 191–196
sick animal, deciding what to do for a, 171–174
urine, monitoring, 167
vaccinations, 174
vital signs, measuring, 168–174
health permits, 295
healthy bovines, signs of, 137
heating box, 251
heavy use area protection (HUAP) zone, 13, 76–78
heel flies, 180, 323
heifer, 24
heifer calf, 24
herd
defined, 24
repositioning entire, 156–157
herd health calendar
breeding season, 176–177
calving season, 175–176
overview, 175
pre-calving season, 175
preweaning season, 177–178
weaning season, 177–178
Hereford/Polled Hereford cattle, 44–45
heritage, 51
heterosis, 41
Highland cattle, 51
high-tensile wire, 58
hock, 26
holding chute, 66, 69–70
holding pen, 66
hook bone, 26
hoop barns, 72
horn bud, 256
horned cattle, 38–39
horns
choosing a breed based on, 38–39
removing, 256–257
hot and humid weather, steps for dealing with, 183
hot carcass weight, 301

hot iron dehorning, 256–257
Houston Livestock Show and Rodeo, 279
HUAP (heavy use area protection) zone, 13, 76–78
hybrid vigor, 41
hypocalcemia (milk fever), 239
hypothermia, 250–251

• I •

IBR (infectious bovine rhinotracheitis), 194, 220
identifying calves and newborns, 243–245
in heat, 150
individual animals, moving, 156
individual cut, marketing your meat by, 298–300
infections in newborns, deterring, 252–253
injections, administering, 185–187
innate behavior of cattle, utilizing, 154–160
insecticide ear tags, 204–205
inspection permits, 295
insurance, 21, 296
internal (subdivision) fences
  electric fencing, 62–65
  gates, planning for, 60–62
  overview, 56, 60
  polywire conductor fences, 60
  semipermanent fences, 60
  types of, 60
internal parasites, 179, 209–210
Internet, buying cattle from, 136
intramuscular injections, 187
intravenous injections, 187
iodine, 86, 88
iron, 86, 88

• J •

jackpot or preview shows, 277–278
Johne's disease, 202–203
joint ill, 252
junior shows, 277

• K •

Kentucky bluegrass, 110
kill floor, 303
killed vaccines, 174

• L •

lactation, 240
Ladies Man moo, 153
land
  budgeting for, 17
  improvement as benefit of raising beef cattle, 11
leasing bulls, 229–230
legalities of selling beef
  insurance coverage, obtaining proper, 296
  permits, obtaining, 295–296
  processing facilities, government rules for, 294–295
legumes, 109–110
leptospirosis, 200, 219
liability insurance, 21
lice, 180–181, 207
limit feeding, 100–101
Limousin cattle, 48
lips, curling, 326
loading chute, 66
local extension agents, 18–19
location liability coverage, 296
loin, 26
low heritability traits, 41
low-stress environment, providing a, 13–14

• M •

magnesium, 85, 88
Mag-Ox (magnesium oxide), 85
Maine-Anjou cattle, 48
major livestock exhibitions, 279
male reproductive anatomy, 29–30

managing your beef business
butchers, finding and working with, 302–306
freezer beef, providing, 297–298
individual cut, marketing your meat by, 298–300
legalities of selling beef, 293–296
prices, setting, 300–302
promotion and marketing, 308–310
seasonal changes to product lineup, 302
selling on the hoof, 297
selling options, 306–308
manganese, 86, 88
mange, 181
manure management
composting, 116–118
feces, monitoring, 167–168
overview, 115–116
spreading manure, 117–118
squirting projectile manure, 322
manure spreader, 117
marbling, 272
marketing, 309–310, 316
mastitis, 240
mature body size, 39
mechanical calf puller, 231–232
meconium, 246
mentor, finding a, 17–18
MidWest Plan Service, 73
milk fever (hypocalcemia), 239
milk production, 40
minerals, 84–88. *See also specific minerals*
mistakes to avoid
appearance, purchasing cattle based only on, 317
bargain cattle, buying, 316–317
calving ease bull, forgoing, 315
complicated, making raising cattle, 319
customer expectations, failing to set proper, 318–319
exhibitions, lack of training before, 318
fences, tolerating weak perimeter, 313–314
forage, running out of, 314
marketing plan, lack of a, 316
water supply, neglecting to keep adequate backup, 315–316
mites, 181, 208
mobile processing units, 295
modified live vaccines (MLVs), 174
moldy feed, 210
mooing noises, understanding, 153–154
moving cattle by using flight zone, 155–158
municipal supply or well as water source, 104–105
muzzle, 26

• N •

National Western Stock Show, 279
navel, disinfecting, 243
navel ill, 251–252
neighbors, maintaining good relationships with, 21–22
new pasture, preparing a, 113–114
newborns. *See also* calves
body temperature, raising, 251
bottle raising, 248–249
breathing difficulties, helping with, 242–243
colostrum, ensuring calf receives enough, 246–248
feeding program for, 96
hypothermia, 250–251
identifying, 243–245
infections, deterring, 252–253
joint ill, 252
navel, disinfecting, 243
navel ill, 251–252
protein recommended for, amount of, 81
scours, preventing and treating, 252–254
vital signs, 250
newspaper as bedding material, 76
nitrogen, analyzing and improving, 115
noises made by cattle, understanding, 153–154
nonprotein nitrogen (NPN), 81
nontraditional feedstuffs, 95
North American International Livestock Exposition, 279
nose rings, 161
nursing calves, protein recommended for, 81

Nutrient Requirements of Beef Cattle (National Academies Press), 100
nutritional needs
carbohydrates, 83–84
fats, 84
minerals, 84–88
overview, 12
plant protein supplements, 84
protein, 80–83
seasonal, 181
vitamins, 89–91
water, 91–92

• O •

offset electrical wire, 58
old grass, ignoring fresh green grass for, 321–322
older, weaned calves, purchase price of, 15
older cows, supporting, 219
omasum, 28
100 percent grass-fed beef
customers for, 272–273
feedsfuffs for, supplying, 273
overview, 271–272
as selling terminology, 309
taste of, 272
type of cattle suited for, 273–274
on-farm processing facilities, 295
online selling options, 291, 307
open cow or heifer, 24
open shows, 277
orchardgrass, 111
Oreo Cookie cattle, 51
organic, 309
ovary, 31
overstocking, 119
oviduct, 31

• P •

packages and labels, options for, 304–305
palpation used to determine pregnancy, 221
paralysis, temporary, 239–240
parasites
controlling, 178–181
external, 204–208
internal, 209–210
pastern, 26
pasteurella, 196
pasture, 93–94
pasture management
absentee owners, using extra pastures to tend cows for, 268–271
analyzing soil samples, 114–115
custom grazing, using extra pastures for, 268–271
evaluating your pasture, 112
fertilizing soil, 114–115
forages, 108–111
galloping around, 323
manure management, 115–118
matching number of cattle to amount of available pasture, 119–120
new pasture, preparing a, 113–114
nitrogen, analyzing and improving, 115
100 percent grass-fed beef, using extra pastures for raising, 271–27
pasture resources needing to match with number of cows, 269–270
pH, analyzing and improving, 115
phosphorus, analyzing and improving, 115
potassium, analyzing and improving, 115
problems with pasture, solutions for, 113
productivity, optimizing, 118–124
rotational grazing, 121–122
stocker calves, using extra pastures for, 263–268
yearlong forage chain, creating, 122–124
yields, estimating pasture, 120
pasture walk, 112
pasture-raised, 309
patch grazing, 119
perennial ryegrass, 111
performance measurements, 139–140
performance test, 290–291
permanent fences
barbed wire fencing, 59
overview, 56
woven wire fencing, 56–59
permits, obtaining, 295–296

personal insurance, 21
pH, analyzing and improving, 115
phosphorus, 85, 88, 115
photographing your cattle, 292
physical injuries, preventing, 19–20
PI-3 (parainfluenza), 195
pin bone, 27
pinkeye, 181, 201–202
plant canopy, 112
plant protein supplements, 84
pneumonia, 192
point of balance, 155
poisonous plants, 210
pole barns, 71
poll, 27
polled cattle, 38
polywire conductor fences, 60
pond, stream, or spring as water sources, 105
postpartum, 237
posts, 58
potassium, 85, 88, 115
pre-calving season, 175
preconditioned calves, 130
predator-prey relationship, 154–155
pre-delivery, 233
preferences, recognizing cattle, 148
pregnancy. *See also* delivery
  abortions, 237–238
  by artificial insemination, 225–229
  BCS (body condition scores), 214–216
  due date, determining, 220–221
  EPD (expected progeny difference), 218
  feeding program for, 99
  first-calf heifers, preparation for, 217–218
  milk fever (hypocalcemia), 239
  older cows, supporting, 219
  palpation used to determine, 221
  paralysis, temporary, 239–240
  preparing cow for, 213–221
  vaccination guidelines, 219–220
presenting your animals and yourself in the best possible way at exhibitions, 285
preventive healthcare, 16
preweaning season, 177–178
prices
  for freezer beef, 298
  setting, 300–302
primping, 286–287
private treaty deals, 289–290
processing facilities, government rules for, 294–295
product liability coverage, 296
productivity, optimizing, 118–124
promotion and marketing, 308–310
protecting your water resources, 124–126
proteins, 80–83
pulse, measuring, 170
purebred cattle, 40–41

• Q •

quality grade, 309
quarantining new cattle, 144

• R •

rate and efficiency of gain, 40
Red Angus cattle, 45–46
red clover, 110
Red Poll cattle, 51
registration papers, 41
replacement heifers. *See* first-calf heifer
reproductive diseases, 199–201
respiration, observing, 171
respiration rate, 171
respiratory diseases
  antibiotic therapies for, 193
  BRSV (bovine respiratory syncytial virus), 195
  BVD (bovine virus diarrhea), 194–195
  diptheria, 192
  H. somnus (haemophilus somnus), 196
  IBR (infectious bovine rhinotracheitis), 194
  overview, 191–193
  pasteurella, 196
  PI-3 (parainfluenza), 195
  pneumonia, 192
  stress, minimizing, 192
  supportive care for, 193
  upper respiratory tract infections, 192
  vaccination plan for, 193

restaurants as selling option, 307–308
reticulum, 28
ribs, 27
rickets, 85
riding (mounting) cattle, 150
ringworm, 207–208
roan, 46
roan markings, 37
rod, 58
rope halter, 281
rotational grazing, 13, 121–122
Royer Farm, choosing cattle for, 52
rubbing behavior, 325
rumen, 28, 102
ruminants, 27

• S •

safety gate, 68
safety issues
  electric fencing, 63–64
  handling facilities, 68
  physical injuries, preventing, 19–20
  shelters, 71
safety pass, 68
safety zone, 155–158
sale barns
  buying cattle from, 135–136
  as source for bottle calves, 129
Saler cattle, 48–49
sales opportunities
  consignment sales, 290
  direct off farm, 306
  farmers' markets, 306–307
  online, 291, 307
  performance test, 290–291
  photographing your cattle for, 292
  private treaty deals, 289–290
  restaurants, 307–308
salt, 86, 88
Santa Gertrudis cattle, 51
schedule for feeding, 100–101
scours, 203, 252–254
scratching cattle to bond, 326
scurs, 38
secondary sex organs, 29
selecting cattle
  bottle calves, 128–130
  bred heifers or cows, 131–132
  bulls, 221–223
  show calves, 130–131
  stocker calves, 265–266
  weaned calves, 130
selenium, 87, 88
self-sufficiency as benefit of raising beef cattle, 10
selling on the hoof, 297
semipermanent fences, 60
separation weaning, 260
sericea lespedeza, 110
shade sources, 73–74
shampooing, 286
shelters
  bedding materials, choosing, 75–76
  cleaning, 71
  drainage for, 70
  existing structures, repurposing, 72–73
  hoop barns, 72
  overview, 70–71
  pole barns, 71
  safety issues, 71
  shade sources, 73–74
  size of, 70
  types of, 71–73
  ventilation for, 70
  wind breaks, 74–75
Shorthorn cattle, 46
show calves, selecting, 130–131
show ring protocol, following, 287–288
show stick, using a, 280
showing your cattle. *See* exhibitions
shrink, 134
sick animal, deciding what to do for a, 171–174
Simbrah cattle, 51
Simmental cattle, 49
sire, 24
smooth bromegrass, 111
soil samples, analyzing, 114–115
spot grazing, 119
spring and summer grazing, 264
spring of rib, 27
squeeze chute, 161
standing heat, 150
state fairs, 279
state-inspected processing facilities, 294
steer, 25
steer calves, 150

stifle, 27
stocker calves
  buyers for, 267
  facilities, setting up, 268
  feeding program for, 97
  overview, 263–264
  pastures, preparing, 267–268
  preparing for, 267–268
  profitability of, 266
  selecting, 265–266
  sources for, 265
  spring and summer grazing, 264
  weight of, 265–266
  winter grazing, 264
stocking rate, 119
stockpiled forage, 123
stomach, 27–28
straw as bedding material, 75–76
stream crossing, building a, 124–125
stress, minimizing, 192
stretching, 152, 166
stun box, 304
subcutaneous injections, 185–186
subdivision fences
  electric fencing, 62–65
  gates, planning for, 60–62
  overview, 56, 60
  polywire conductor fences, 60
  semipermanent fences, 60
  types of, 60
sulfur, 86, 88
summer, accommodations in, 13
summer grazing, 264
supplements
  mineral, 87–88
  proteins, 82
supplies
  for delivery, 231–232
  for exhibitions, 283–284
  for feeding program, 102–106
support system
  feed dealer, 19
  local extension agents, 18–19
  mentor, finding a, 17–18
  overview, 17
  veterinarian, 18
supportive care for respiratory diseases, 193
surgical castration, 258
sweet clover poisoning, 212
switch, 27

• T •

tailhead, 27
tails, reading body language of, 152
tank heaters, 106
tanks or troughs as water sources, 105–106
taste of 100 percent grass-fed beef, 272
tattoos used for identifying calves, 244–245
tax permits, 295
TB (tuberculosis), 203
TDN (total digestible nutrients), 83–84
tensile strength of woven wire fencing, 56–57
terminology
  for anatomy of cattle, 25–31
  for cattle stomach, 27–28
  for cattle-selling, 308–309
  for female reproductive anatomy, 30–31
  for male reproductive anatomy, 29–30
  for types of cattle, 23–25
terpene, 109
test station, 135, 290–291
testicles, 29
Texas Longhorn cattle, 50
threat display, 162
thyroxin, 86
ticks, 180
tight calf crop, 225
training cattle, 160, 280–283
transporting cattle, 142–143
trichomoniasis, 201, 219
trimming, 286
troughs as water sources, 105–106
tube dehorning, 256
turbidity, 105
twins, 237

• U •

udder, 31
understocking, 119
unhealthy bovines, signs of, 137
upper respiratory tract infections, 192
urine, monitoring, 167
USDA inspected processing facilities, 294
uterus and uterine horns, 31

• V •

vaccination guidelines
  for calves, 255
  overview, 174
  for pregnancy, 219–220
  for respiratory diseases, 193
vagina, 31
veal, 32
veterinarian, 18
vibriosis, 200, 219
vital signs
  measuring, 168–174
  newborns, 250
vitamin A, 89, 181
vitamin D, 89–90
vitamin E, 90
vitamin K, 90
vitamins, 89–91. *See also specific vitamins*
vomit, 101
vulva, 31

• W •

warbles, 180
warm-season annuals, 123
washy forage, 322
water (drinking)
  daily requirements for, 92
  nutritional needs, 91–92
  overview, 91
  proximity to, 91–92
  sources for, 104–106
  trough, not drinking from, 322–323
water management
  backup water supply, neglecting to keep adequate, 315–316
  farm ponds, 125–126
  flash grazing, 126
  protecting your water resources, 124–126
  stream crossing, building a, 124–125
water sources
  feeding program, 104–106
  municipal supply or well, 104–105
  overview, 12
  pond, stream, or spring, 105
  tanks or troughs, 105–106
weaned calves, selecting, 130
weaning, 25, 258–260
Weaning Cry moo, 153
weaning season, 177–178
weather, keeping cattle comfortable during extreme, 182–183
weight of stocker calves, 265–266
wet-aging, 309
white clover, 109
wind breaks, 74–75
windbreak fences, 74
winter
  accommodations in, 13
  grazing, 264
withdrawal time, 185, 295
woodchips as bedding material, 76
working chute, 66
wounds, treating, 189
woven wire fencing
  components of, 57–58
  end assemblies, 58
  high-tensile wire, 58
  offset electrical wire, 58
  overview, 56–57
  posts, 58
  tensile strength of, 56–57

• Y •

yearlings
  defined, 25
  purchase prices, 15
yearlong forage chain, creating, 122–124
yield grade, 309
yields, estimating pasture, 120

• Z •

Zebu cattle, 43
zinc, 87, 88
zoning regulations, 22
zoonotic diseases, 20–21

Apple & Mac

Pad 2 For Dummies,
3rd Edition
978-1-118-17679-5

Phone 4S For Dummies,
5th Edition
978-1-118-03671-6

Pod touch For Dummies,
3rd Edition
978-1-118-12960-9

Mac OS X Lion
For Dummies
978-1-118-02205-4

Blogging & Social Media

CityVille For Dummies
978-1-118-08337-6

Facebook For Dummies,
4th Edition
978-1-118-09562-1

Mom Blogging
For Dummies
978-1-118-03843-7

Twitter For Dummies,
2nd Edition
978-0-470-76879-2

WordPress For Dummies,
4th Edition
978-1-118-07342-1

Business

Cash Flow For Dummies
978-1-118-01850-7

nvesting For Dummies,
6th Edition
978-0-470-90545-6

Job Searching with Social
Media For Dummies
978-0-470-93072-4

QuickBooks 2012
For Dummies
978-1-118-09120-3

Resumes For Dummies,
6th Edition
978-0-470-87361-8

Starting an Etsy Business
For Dummies
978-0-470-93067-0

Cooking & Entertaining

Cooking Basics
For Dummies, 4th Edition
978-0-470-91388-8

Wine For Dummies,
4th Edition
978-0-470-04579-4

Diet & Nutrition

Kettlebells For Dummies
978-0-470-59929-7

Nutrition For Dummies,
5th Edition
978-0-470-93231-5

Restaurant Calorie Counter
For Dummies,
2nd Edition
978-0-470-64405-8

Digital Photography

Digital SLR Cameras &
Photography For Dummies,
4th Edition
978-1-118-14489-3

Digital SLR Settings
& Shortcuts
For Dummies
978-0-470-91763-3

Photoshop Elements 10
For Dummies
978-1-118-10742-3

Gardening

Gardening Basics
For Dummies
978-0-470-03749-2

Vegetable Gardening
For Dummies,
2nd Edition
978-0-470-49870-5

Green/Sustainable

Raising Chickens
For Dummies
978-0-470-46544-8

Green Cleaning
For Dummies
978-0-470-39106-8

Health

Diabetes For Dummies,
3rd Edition
978-0-470-27086-8

Food Allergies
For Dummies
978-0-470-09584-3

Living Gluten-Free
For Dummies,
2nd Edition
978-0-470-58589-4

Hobbies

Beekeeping
For Dummies,
2nd Edition
978-0-470-43065-1

Chess For Dummies,
3rd Edition
978-1-118-01695-4

Drawing For Dummies,
2nd Edition
978-0-470-61842-4

eBay For Dummies,
7th Edition
978-1-118-09806-6

Knitting For Dummies,
2nd Edition
978-0-470-28747-7

Language & Foreign Language

English Grammar
For Dummies,
2nd Edition
978-0-470-54664-2

French For Dummies,
2nd Edition
978-1-118-00464-7

German For Dummies,
2nd Edition
978-0-470-90101-4

Spanish Essentials
For Dummies
978-0-470-63751-7

Spanish For Dummies,
2nd Edition
978-0-470-87855-2

Available wherever books are sold. For more information or to order direct: U.S. customers visit www.dummies.com or call 1-877-762-2974.
U.K. customers visit www.wileyeurope.com or call (0) 1243 843291. Canadian customers visit www.wiley.ca or call 1-800-567-4797.
Connect with us online at www.facebook.com/fordummies or @fordummies

Math & Science

Algebra I For Dummies, 2nd Edition 978-0-470-55964-2

Biology For Dummies, 2nd Edition 978-0-470-59875-7

Chemistry For Dummies, 2nd Edition 978-1-1180-0730-3

Geometry For Dummies, 2nd Edition 978-0-470-08946-0

Pre-Algebra Essentials For Dummies 978-0-470-61838-7

Microsoft Office

Excel 2010 For Dummies 978-0-470-48953-6

Office 2010 All-in-One For Dummies 978-0-470-49748-7

Office 2011 for Mac For Dummies 978-0-470-87869-9

Word 2010 For Dummies 978-0-470-48772-3

Music

Guitar For Dummies, 2nd Edition 978-0-7645-9904-0

Clarinet For Dummies 978-0-470-58477-4

iPod & iTunes For Dummies, 9th Edition 978-1-118-13060-5

Pets

Cats For Dummies, 2nd Edition 978-0-7645-5275-5

Dogs All-in One For Dummies 978-0470-52978-2

Saltwater Aquariums For Dummies 978-0-470-06805-2

Religion & Inspiration

The Bible For Dummies 978-0-7645-5296-0

Catholicism For Dummies, 2nd Edition 978-1-118-07778-8

Spirituality For Dummies, 2nd Edition 978-0-470-19142-2

Self-Help & Relationships

Happiness For Dummies 978-0-470-28171-0

Overcoming Anxiety For Dummies, 2nd Edition 978-0-470-57441-6

Seniors

Crosswords For Seniors For Dummies 978-0-470-49157-7

iPad 2 For Seniors For Dummies, 3rd Edition 978-1-118-17678-8

Laptops & Tablets For Seniors For Dummies, 2nd Edition 978-1-118-09596-6

Smartphones & Tablets

BlackBerry For Dummies, 5th Edition 978-1-118-10035-6

Droid X2 For Dummies 978-1-118-14864-8

HTC ThunderBolt For Dummies 978-1-118-07601-9

MOTOROLA XOOM For Dummies 978-1-118-08835-7

Sports

Basketball For Dummies, 3rd Edition 978-1-118-07374-2

Football For Dummies, 2nd Edition 978-1-118-01261-1

Golf For Dummies, 4th Edition 978-0-470-88279-5

Test Prep

ACT For Dummies, 5th Edition 978-1-118-01259-8

ASVAB For Dummies, 3rd Edition 978-0-470-63760-9

The GRE Test For Dummies, 7th Edition 978-0-470-00919-2

Police Officer Exam For Dummies 978-0-470-88724-0

Series 7 Exam For Dummies 978-0-470-09932-2

Web Development

HTML, CSS, & XHTML For Dummies, 7th Edition 978-0-470-91659-9

Drupal For Dummies, 2nd Edition 978-1-118-08348-2

Windows 7

Windows 7 For Dummies 978-0-470-49743-2

Windows 7 For Dummies, Book + DVD Bundle 978-0-470-52398-8

Windows 7 All-in-One For Dummies 978-0-470-48763-1

Available wherever books are sold. For more information or to order direct: U.S. customers visit www.dummies.com or call 1-877-762-2974. U.K. customers visit www.wileyeurope.com or call (0) 1243 843291. Canadian customers visit www.wiley.ca or call 1-800-567-4797.

Connect with us online at www.facebook.com/fordummies or @fordummies